Wazzat Bug?

A Guide to Arthropods of Culpeper County and Northern Virginia

Lee Alloway

Ancient Eagle Press
Culpeper, Virginia

ISBN: 978-0-9971923-0-8
Library of Congress Control Number: 2021923529

Copies available at https://ancient-eagle-press.square.site/

To the curious, the gentle, the children of the earth.

ACKNOWLEDGMENTS

I am not an entomologist by trade and could not have produced this book were it not for the outstanding entomologists and photographers who have inspired and taught me more than I thought I wanted to know about bugs. The BugShot team of John and Kendra Abbot, Thomas Shahan, Peter Naskrecki, Nikki Bay, and Alex Wild are first class educators who care deeply about the world and all its creatures. While some bugs are distinctive, most require careful examination and specialized knowledge for authoritative identification. Lacking specialized knowledge, I have relied on various regional web sites and field guides as well as the community resources of Bugguide.net and iNaturalist.org to flesh out this volume. Finally, this book would not have been possible without the patience and forbearance of my wife who is not a fan of bugs. Thanks, Jan, for putting up with the constant stream of “bugs in a jug” that have filled my hours throughout the COVID isolation.

Contents

Introduction

How many bugs are alive and crawling about today? I don't know. Neither do you. Some smart bug-estimator has pegged the figure at ten quintillion. That's 10,000,000,000,000,000,000! About one million species have been identified, and scientists have estimated that the true number of species is between two million and thirty million. Talk about hedging your bets!

For many of us living in urban isolation, the bugs we encounter are flies, mosquitos, ants, gnats, and the occasional cockroach. Once we migrate to the suburbs, our world expands to include ticks, butterflies, wasps, and a chorus of unseen creatures singing in the night. It's not that there are only a few insects around. It's just that we filter them out of our consciousness. But look around carefully and you will see...

SCARY THINGS! We see a bug and our first reaction is to scream, swat, stomp or at least shoo the creature away. We fear for our safety, sure that the bug who weighs less than a dust ball is somehow set on our destruction, armed with mighty jaws, lethal poisons and a stinger aimed at our tender flesh. We stand no chance against the cunning of a bug who wants to suck our blood and leave us with an annoying itch or an incurable disease.

BUT WAIT! Before you bring that newspaper down or stomp that bug into oblivion, isn't there just a moment when you wonder what the bug really is? Which one of those many species is it, and what is it doing in your back yard? Is it really a threat to you, or is it perhaps out there working on your behalf?

In this moment of moral ambiguity, grab a plastic cup, capture the invader, and spend some time studying it. Look at the colors, the amazing eyes, the complex legs. If you have a field guide handy, you will find hundreds of bug pictures, many of which look a bit like your bug, none of which look exactly like it. If you turn to the Internet, your choices are almost limitless, offering hours of entertainment before you discover the bug you thought looks most like yours happens to live only in the Andes above 14,000 feet.

That's why you need WAZZAT BUG! After years of extensive research and photography in suburban back yards and watery areas of Virginia, I have limited this book to just bugs you may actually find in your yard. I know this because they are bugs I have actually found in my back yard.[1] So look around, grab a plastic cup, go find a bug and ID the critter before you let it go about its business.

A million more: With millions of bugs to choose from there is a high likelihood you will find bugs not mentioned in this book. At the end of the book, I have listed several excellent references to help you identify those.

---

[1] Unless noted otherwise, all the bugs in this book were photographed in Culpeper, Orange, or Fairfax counties. All were photographed alive where observed. Some were additionally photographed in a white box before being returned to their natural environment. Except ticks. I have my standards.

Definitions

First let's agree on some definitions:

Bug: For purposes of this field guide, a "bug" will be any small creature with six or eight legs in its adult form. Most bugs will be insects or arachnids (spiders and their kin,) although I've included some collembola as well as woodlice (who have lots of legs) because there are so many of them in your yard and they look a lot like bugs.

Insect: A bug that has

- Three pairs of legs
- Three body parts: head, thorax, and abdomen
- Typically, two pair of wings[2];
- One pair of antennae
- Usually, one pair of compound eyes

Almost one million species of insects have been identified, with millions more still to be described. Beetles alone count for almost 400,000 described species, with an estimated one million more species crawling around undescribed.

Arachnid: Spiders and their kin having

- One or two body parts
- Four pairs of legs
- No antennae

There are around 35,000 known species of spiders in the world, with 30,000 of them present in the United States. There are many more that haven't been identified yet.

---

[2] In some families one pair of wings has evolved into something else. In diptera (flies) one pair of wings has become a stabilizing structure called a haltere. In beetles, the forewing has hardened into a protective cover called an elytron.

Harmful/beneficial: Bugs are not inherently harmful or beneficial. They are simply doing what they are designed to do. For the most part, we don't see the bugs around us and we have no opinion one way or the other about "harmful" or "beneficial." It is only when the normal function of the bug intersects our own lives that we pass judgement. If we have a vegetable garden we may be competing with "harmful" insects for crops; raising flowers, we may find insect damage reduces the commercial or esthetic value of our product. A wasp may be "beneficial" when it kills caterpillars that eat your flowers, but "harmful" when you step on it and it stings you.

In many cases we experience "harmful" bug populations as an unintended consequence of our own action. We may introduce a new animal into an environment that has no offsetting predator. We may use insecticide or herbicide to modify our environment, and inadvertently kill off "beneficial" insects and enable previously benign insect populations to become "harmful." In the short term, an abundance of prey may cause the population of predators to swell, which leads to overconsumption of prey and a collapse of the predator population. Over the long term, an equilibrium exists in nature. Predators and prey are in balance[3].

Icky: There are no icky bugs[4]

---

[3] As a general rule, the best approach to controlling bugs is to let the bugs regulate themselves. As a practical matter, life is a compromise. I leave the outdoor bugs alone, but anything crawling or flying inside the house has a short life expectancy. Even outside there are times intervention is needed. Yellow jackets, for instance, are territorial and can be aggressive. They should not be nesting in an area where children or pets will be playing and are best away from your work and gardening areas. If you have water features in the yard, it may be necessary to control mosquitos with targeted biochemical products.

[4] There is some disagreement on this point.

Organization

For thousands of years people have been studying and classifying bugs. The best known system of classification (taxonomy) is the binomial nomenclature devised by Carl Linnaeus in the mid-1700s. While Linnaeus' system describes three Kingdoms of nature (Plant, Animal, and Mineral), we will only be concerned with animals.

Within the animal Kingdom, individuals are grouped into

- Phyla[5] (singular "Phylum") which contain
    - Orders (singular "Order") which contain
        - Families (singular "Family") which contain
            - Genera (singular "Genus") which contain
                - Species (singular "Species").

The "binomial nomenclature" refers to the Genus and Species of an individual. A monarch butterfly for instance is in the Animal kingdom, Phylum *Arthropoda*, Class *Insecta*, Order *Lepidoptera*, Family *Nymphalidae*, Genus *Danaus*, and Species *plexippus*, so its binomial nomenclature is *Danaus plexippus.* When written, it is generally referred to as *"Danaus plexippus"* when it first appears but is subsequently noted simply as "*D. plexippus*." I have occasionally used that convention in this book. When the species of a bug is in doubt, "sp." will serve as a placeholder for "species."

While Linnaeus got things started, there are now more than one million organisms that have been classified. To prevent overcrowding, taxonomists have created numerous subdivisions such as "subphyla," "superorder," "suborder", "subfamily," etc., but the basic structure remains intact. The system has proven to be flexible enough to accommodate renaming and shifting of individuals within the system as more is understood about the true nature and relationships of the individuals. Increased use of genetic sequencing has led to frequent

---

[5] All the bugs in this book belong in the phylum Arthropoda.

reclassification of insects and spiders. Classifications used in the volume are current as of 2021 but are all subject to revision as scientists learn more.

This volume has been organized to keep similar bugs together. Major categories correspond to taxonomic Orders: Arachnids (spiders and their relatives), Insects, and Other Bugs (collembola and woodlice). Within these major groups, bugs are grouped in taxonomic Families[6]. Finally, they are listed alphabetically by genus and species. The index has a list of bugs by scientific name and by common name where those are available.

With that, let's get started!

Size matters. Most insects are less than 1" long so they are generally measured in millimeters (mm.) and I use that convention in this book. There are 26 mm. in an inch. Images in this book are not to scale.

Speaking of size, when size is given in this book it refers to the body length, not including legs. The exception is moths and butterflies. Sizes there refer to wingspan except as noted.

---

[6] You will find a few exceptions to this where similar bugs are grouped together despite being of different families.

Class *Arachnida* (Arachnids) - Spiders and Kin

Order Araneae (Spiders)

Spiders! Eek, spiders! OK, let's get the scary part over with. Nobody likes spiders. You don't like them in your yard, and you certainly don't like them in your house. They're creepy and hairy and their webs get caught on your face and they have fangs and they live in haunted houses.... There's just no end to the bad press these bugs get. The *Araneae* are bugs with few admirers, yet they are some of the most beautiful and useful insects around. For the most part they stay out of sight. Many prefer to hunt in the safety of darkness. Others may be easy to find as they stalk their prey in the daylight, relying on sharp eyesight and quick reflexes to avoid predators. While all spiders have fangs and most have venom, the number of spiders that can harm a human is extremely low[7]. In the eastern United States, the number is 1: The Black Widow (5 species, of which *Latrodectus mactans* is most common).[8] As scary as the Black Widow can be, you really have to annoy this spider to get bitten. Most bites occur when you accidentally step on or grab one. Of the 2500 black widow bites reported in the U.S. each year, none have proven fatal. In fact, in the 20th Century, only about 100 fatal spider bites were reported in the entire world: One per year, world-wide. By contrast there were 432 U.S. deaths by lightning strikes in 1942. You're much safer with the spiders than with the storms. Still, the black widow is not a spider to trifle

---

[7] While most spider toxin is not medically significant for humans, some individuals may have allergic reactions or local site irritation from spider bites. Be respectful of spiders. Watch, study, take pictures of the spider, but if you are in doubt about the identification of the spider and how to handle it properly, it is best to leave it alone.

[8] The **brown recluse** spider, *Loxosceles reclusa*, can be found from Kentucky to Texas, but has not established itself on the East Coast...yet. Reports of brown recluse bites on the east coast might be cases of mis-identification or spiders being transported from their western habitat to the east in produce or packing material. However, it is likely the brown recluse will take up residence in the area eventually if it hasn't already.

with. The other east coast spiders may look scary, but look closer and you're in for a treat! They are a fascinating group.

ID Tip: You can tell the family of spiders by looking at their eyes. Almost all spiders have 8 eyes in a patten distinctive to each family. (About 1% of spiders have 6 eyes; a few outliers have 4, 2 or no eyes). In this guide, the eye pattern of spiders is displayed under the heading for each section.

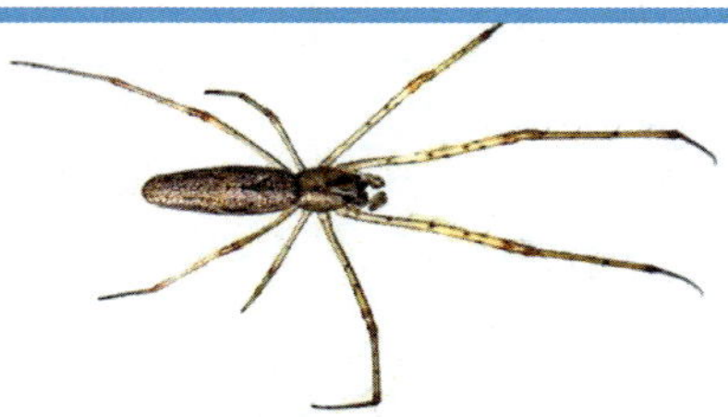

ID Tip #2: Now that you've got the spider in the right family, identification becomes more difficult. While some spiders are unique and easy to ID, many spiders change appearance as they grow and molt; many species exhibit extreme sexual dimorphism, usually with males being much smaller and nondescript. Even extravagantly patterned and colored spiders of the same species can have very different patterns and colors. Species identification may require a microscope to count the number of hairs above an eye or the relative size of leg segments. Sometimes Family is as close as you can get.

Family Agelenidae – Grass Spiders

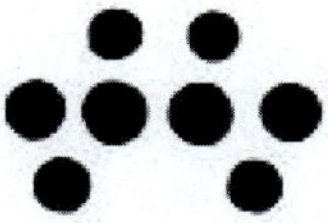

Agelenidae Eye Pattern

Agelenopsis sp., commonly known as the American Grass Spider, is a funnel weaver. There are 14 species of Agelenopsis in our area. The larger specimens (depending on species) can grow to about 19 mm in body length. They have two prominent hind spinnerets and two dark bands running down either side of the cephalothorax.

Agelenopsis builds a flat non-sticky horizontal web near the ground with a barrier web above and a funnel web in which to hide. When a passing insect hits the barrier, it bounces into the horizontal web and gets tangled long enough for the spider to rush out and envenomate it, quickly dragging it back into the safety of the funnel.

American Grass Spider
(*Agelenopsis sp.*)
12 mm.

Agelenopsis sp.
10 mm.

Family Amaurobiidae – Hackmesh Weavers

Amaurobiidae Eye Pattern

Hackmesh Weaver
(*Amaurobius ferox*)
20 mm.

The Hackmesh Weaver is also known as the Black Lace Weaver. This individual was about 20mm, as large as any Hackmesh Weaver. These spiders have no venom glands. Neither do they build sticky webs. So how do they catch food? They build tangled webs of fine silk that trap their prey. The spider wraps the prey and marinades it in digestive juices, then sucks the dissolved prey out of the wrappings. Sweet.

Family Anyphaenidae – Ghost Spiders

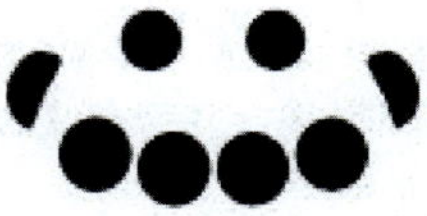

Anyphaenidae Eye Pattern

Anyphaenidae build silken retreats, or sacs, usually on plant terminals, between leaves, under bark or under rocks. There are more than 500 species in over 50 genera worldwide. These are important predators in several agricultural systems, especially tree crops. They can detect and feed on insect eggs, despite their poor eyesight.

Garden Ghost Spider
(Hibana gracilis)
10 mm.

1 mm. Ghost Spiderling

Wulfila albens
3 mm.

Wulfila saltabundus
4 mm.

Family Araneideae -- Orbweavers

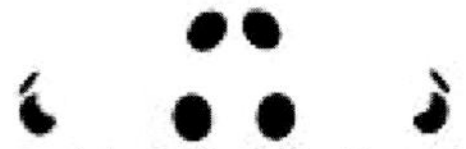

Araneidae Eye Pattern

Orb-weaver spiders are the most common group of builders of spiral wheel-shaped webs often found in gardens, fields, and forests. Araneidae is the third-largest family of spiders (behind jumping spiders and sheet weavers). Araneid webs are constructed with a framework of non-sticky silk covered in sticky droplets. Many orb-weavers build a new web each day. Towards evening, the spider will consume the old web and spin a new web in the same general location. Some orb-weavers construct a crisscross band of silk (stabilimentum) through the center of the web. This is the habit of *Argiope* – the yellow and banded garden spiders found throughout North America and one of the most visible in Virginia gardens.

Female Araneidae are often large and colorful, spinning webs that sometimes reach three-feet across. Male Araneidae are smaller and are not permanent web residents but may wander from web to web looking for a mate. I have observed several males on a web.

Spider webs are an interesting study in themselves. Most spiders can produce different kinds of silk and construct webs with silks to suit different requirements. You can study spider silk on-line or in books, but here's a more literary introduction:

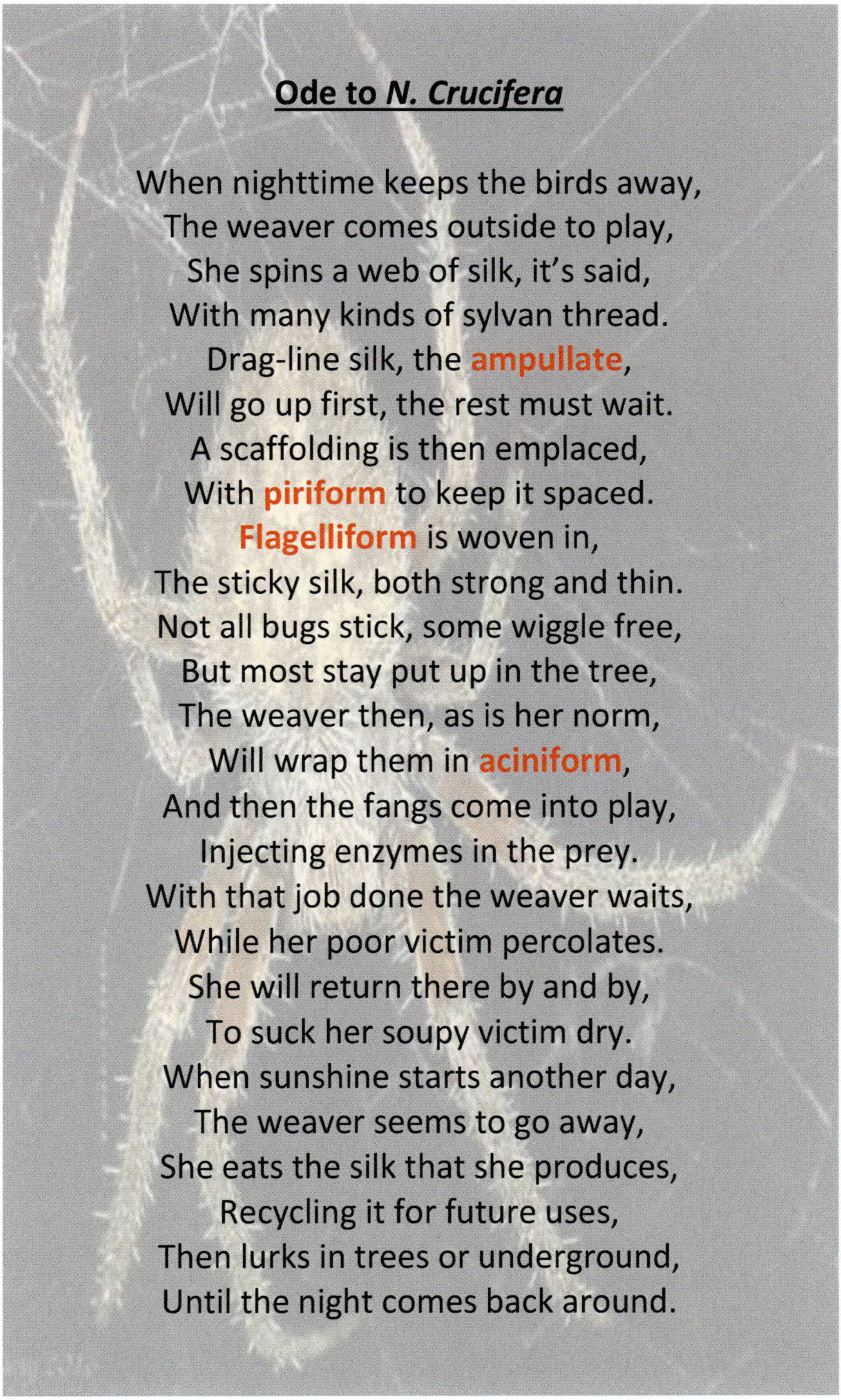

<u>Ode to *N. Crucifera*</u>

When nighttime keeps the birds away,
The weaver comes outside to play,
She spins a web of silk, it's said,
With many kinds of sylvan thread.
Drag-line silk, the **ampullate**,
Will go up first, the rest must wait.
A scaffolding is then emplaced,
With **piriform** to keep it spaced.
Flagelliform is woven in,
The sticky silk, both strong and thin.
Not all bugs stick, some wiggle free,
But most stay put up in the tree,
The weaver then, as is her norm,
Will wrap them in **aciniform**,
And then the fangs come into play,
Injecting enzymes in the prey.
With that job done the weaver waits,
While her poor victim percolates.
She will return there by and by,
To suck her soupy victim dry.
When sunshine starts another day,
The weaver seems to go away,
She eats the silk that she produces,
Recycling it for future uses,
Then lurks in trees or underground,
Until the night comes back around.

Acacesia hamata

7 mm.

The only Acacesia species found in North America. It is distinguished by the "dagger" outline on its back. It is generally found in mid-summer.

Acanthepeira cherokee

9 mm.

Acanthepeira sp

4 mm.

Probably *A. stellate*, the Starbellied Orbweaver. Lives close to the ground, generally no higher than 4 feet. Constructs a small vertical web to capture prey. May glue a small brown egg sac to a stem or leaf.

Argiope aurantia has several common names: Yellow garden spider, black and yellow garden spider, golden garden spider, writing spider, zigzag spider, hay spider, corn spider, or McKinley spider. The variety of common names speaks to the wide distribution, large size and visibility of this species. Males range from 5–9 mm; females range from 19–28 mm. Generally, the female is visible on the web passively hunting or actively eating. The smaller male may be present but is not a permanent resident. The web of the yellow garden spider is a distinctive circular shape up to 2 feet in diameter, with a dense zigzag of silk, known as a stabilimentum, in the center. This spider may bite if disturbed or harassed, but the venom is harmless to non-allergic humans, roughly equivalent to a bumblebee sting.

Argiope aurantia (female)
25 mm.

Argiope aurantia (male)
9 mm.

Araneus cingulatus
4 mm.
Sometimes called the red-spotted Orbweaver; a small spider most often sighted mid-summer.

Araneus guttulatus
3 mm.
Another small but colorful spider most often seen in the summer.

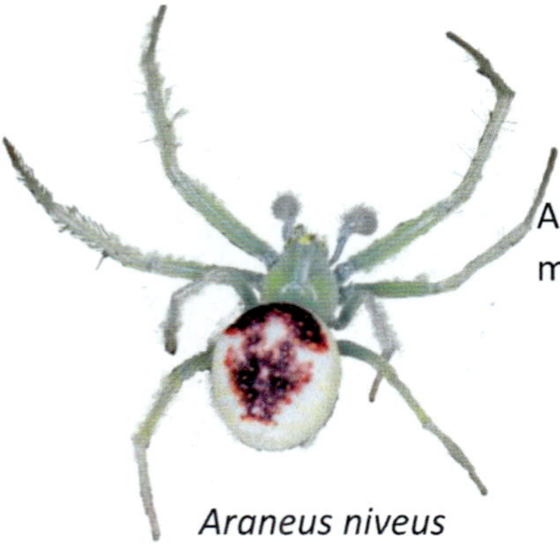

Araneus niveus
3 mm.

Marbled Orbweaver
(*Araneus marmoreus)*
14 mm.
Also known as the pumpkin spider, *A. marmoreus* comes in a dizzying variety of colors, but this yellow variety is considered typical.

Six-spotted Orbweaver
(Araniella displicat)
4 mm.
"Six-spotted" refers to six black dot (three on each side), barely visible in this picture.

The genus *Cyclosa* are Trashline Orbweavers, so designated because they decorate their small webs with a line of insect parts from their prey as well as other bits of detritus that may have fallen into the web. The spider itself is usually in the middle of the line of "trash," blending in perfectly.

Cyclosa turbinata
6 mm.

Most often seen in late spring, *Eustala* thrive in a variety of habitats from open fields to marshland

Eustala sp .
6 mm.

Furrowed Orbweaver
(*Larinioides cornutus*)
aka the Foliate Spider
10 mm.

Larinioides sp

Labyrinth Orbweaver
(*Metapeira labyrinthea*)
6 mm.

Of the 100+ species in the genus *Micrathena* (Spiny Orbweavers) only three live in the eastern U.S.

Spined Micrathena
(*Micrathena gracilis*)
4-10 mm.
Usually seen hanging upside down on a web line.

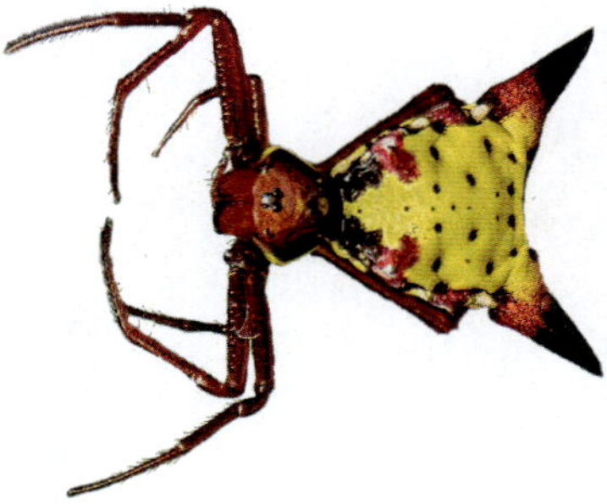

Arrowshaped Micrathena (F)
(*Micrathena sagittate*)
8 mm. (F) / 4 mm. (M)

Of the eight species of genus *Neoscona* (Spotted Orbweavers) found in the U.S., I have seen three in the local area:

Hentz Orbweaver
(*Neoscona crucifera*)
12 mm.

Arabesque Orbweaver
(*Neoscona arabesca*)
11 mm.

Red-femured Spotted Orbweaver
(*Neoscona domiciliorum*)
12 mm.

Arrowhead Orbweaver (F)
(*Verrucosa arenata*)
7-14 mm.

While most orbweavers remain heads-down on the web, *V. grenata* is usually head-up.

*Leucauge venusta***,** the Orchard Spider, is quite common, quite colorful and rather small. Although only about 4-6 mm, its shiny surface and colorful patterns make is easy to find in most environments. Most have a small red dot on the underside of the abdomen (ala the black widow) but this spider is harmless to humans. Unlike most orbweavers, *L. venusta* builds horizontal webs and spends its time hanging from the middle.

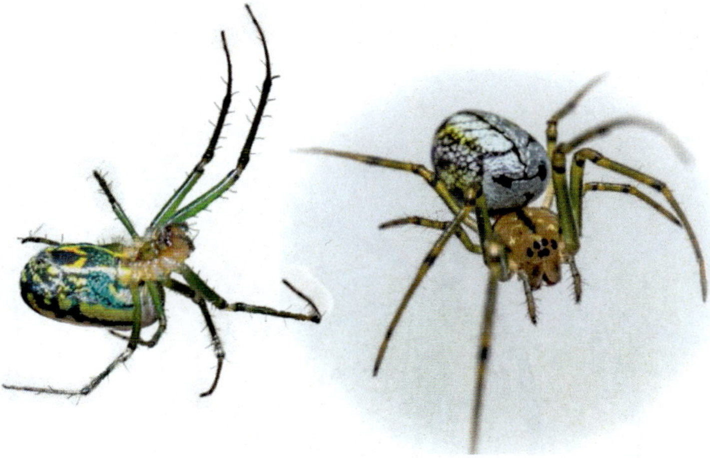

Orchard Spider
(*Leucauge venusta)*
6 mm.

Family Clubionidae – Sac Spiders

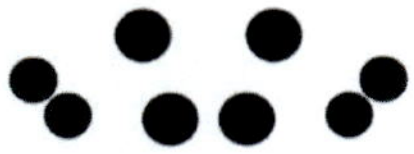

Clubionidae Eye Pattern

Sac spiders do not build webs but can quickly spin a protective shelter resembling a cocoon, hence the name "sac spider." When out of the sac they actively hunt insects and other spiders.

Leafcurling Sac Spider
(*Clubiona sp.*)
8 mm.

White Sac Spider
(Elaver excepta)
8 mm

Of the eight species of *Elvater* in the U.S., only *Elvater excepta* is found on the East Coast.

Family Dysderidae – Woodlouse Hunters

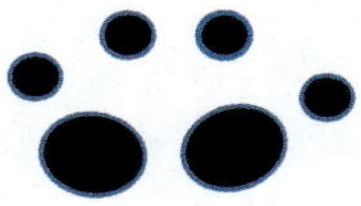

Dysderida Eye Pattern

Dysderida species are native to Central Asia, Europe and Northern Africa. One species, D. crocata, is widespread in the United States.

Woodlouse Spider
(*Dysdera crocata)*
5 mm.

AKA sowbug killer or European garden spider; one of the few spiders known to hunt woodlice. Some species grow to 20 mm and can inflict a painful (though not poisonous) bite.

Family Gnaphosidae – Ground Spiders

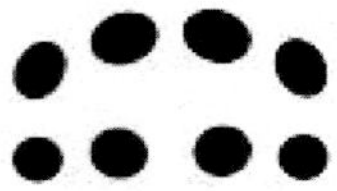

Gnaphosidae Eye Pattern

Gnaphosidae are Ground Spiders, the seventh largest spider family with nearly 2,000 described species distributed worldwide. Ground Spiders generally have barrel-shaped anterior spinnerets that are one spinneret diameter apart. All Ground Spiders lack a prey-capture web and generally run prey down on the surface.

Eastern Parson Spider
(Herpyllus ecclesiasticus)
6 mm. (male), 8-11 mm. (female)

Family Linyphiidae -- Sheetweb Spiders

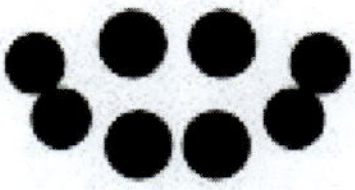

Linyphiidae Eye Pattern

650 species have been identified in North America, but they are all small and not well studied, so it is likely many more species are yet to be discovered. Appropriately, these are of the Erigoninae (dwarf spider) subfamily of Linyphiidae sheetweb spiders.

Dwarf Spider <2 mm.

Black-tailed Red Sheetweaver
(*Florinda coccinea*)
3 mm.

Bowl and Doily Weaver
(*Frontinella pyramitela*)
3 mm.

Linyphiid sp .
3 mm.

Family Lycosidae – Wolf Spiders

Lycosidae Eye Pattern

Wolf Spiders are agile hunters with excellent eyesight. They live mostly in solitude, hunt alone, and do not spin webs. Some are opportunistic hunters pouncing upon prey as they find it or even chasing it over short distances. Their excellent eyesight is enhanced by retroreflective tissue in their four nonsmall eyes. Flashing a beam of light over the spider produces eyeshine easily observed at night.

Wolf Spiders are unique in the way that they carry their eggs. The egg sac is attached to the spinnerets at the end of the abdomen, allowing the spider to carry her unborn young with her. When the spiderlings emerge from their protective silken case, they climb their mother's legs and settle atop her abdomen. The mother carries the spiderlings for several weeks before they are large enough to disperse and fend for themselves. A large spider that appears fluffy might be a wolf spider with a back full of kids.

Allocosa sp
7 mm.

Avid Split Wolf Spider
(*Schizocosa avida*)
12 mm.

Tigrosa georgicola
10-17mm. (male), 16-22mm (females)

Tigrosa helluo
10-12mm. (male)
18-21 mm. (female)

Pardosa sp (*milvina/moesta/modica*)
8 mm.
a Thin-legged Wolf Spider

Rabid Wolf Spider
(*Rabidosa rabida*)
11-12mm. (male)
16-21 mm. (female)

Family Oxyopidae – Lynx Spiders

Oxyopidae Eye Pattern

Lynx Spiders are ambush predators with characteristic spines on their legs. Often found on flowers, they are relatively shy spiders likely to hide from approaching humans.

Striped lynx Spider (M)
(*Oxyopes salticus)*
4-6 mm.

Frequently found in weedy or leafy environments, *O. salticus* has been noted as beneficial in control of agricultural pests. However, it is indiscriminate and will hunt bees and other pollinators as well.

Family Pholcidae – Cellar Spiders

Pholcidae Eye Pattern

The Pholcidae family, commonly known as Cellar Spider, daddy long-legs spider, granddaddy long-legs spider, carpenter spider, daddy long-legger, vibrating spider and skull spider, includes over 1800 species. Pholcids are thin and fragile arachnids. The body (resembling the shape of a peanut) is approximately 2–10 mm long, and the legs may be up to 50 mm long. In areas of human habitation pholcids construct webs in undisturbed areas in buildings such as attics and cellars (hence the common name "cellar spider").

Pholcus manueli
5 mm.

Family Pisauridae - Nursery Web Spiders

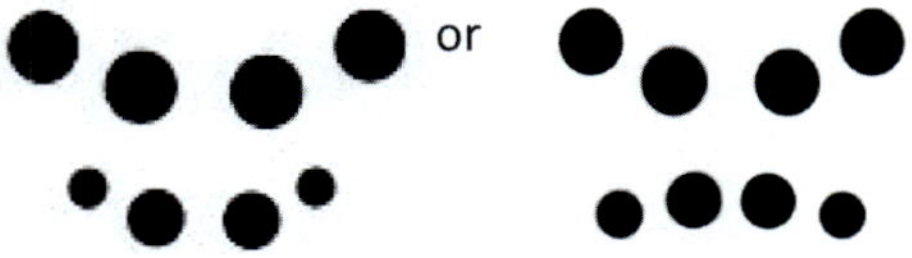

Pisauridae Eye Pattern

Nursery web spiders are similar to wolf spiders (Lycosidae) except they typically rest with legs flattened, carry their egg sacs on their jaws or pedipalps (instead of attaching them to their spinnerets), and all eight eyes are approximately the same size.

Nursery web spiders in the genus *Dolomedes* are covered in short, velvety hairs which are hydrophobic. This allows them to stand or run on the water. *Dolomedes* can also climb beneath the water, with air trapped in the body hairs. They breathe with book lungs beneath their abdomens, using the trapped air to breathe while submerged.

Six-spotted Fishing Spider
(*Dolomedes triton)*
10-20 mm. body

White-banded Fishing Spider
(*Dolomedes albineus*)
24 mm. body – 80 mm. across

Dark Fishing Spider
(*Dolomedes tenebrosus*)
24 mm. body – 80 mm. across

Family Salticidae – Jumping Spiders

Salticidae Eye Pattern

The Jumping Spiders are the largest spider family with more than 6,000 species worldwide, about 13% of all described spider species. They are certainly abundant in Virginia. In addition to being one of the most beneficial bugs in your garden, the jumpers are one of the few bugs that seem to have a personality and an interest in interacting with you. With two large eyes and three additional pairs of smaller eyes, they have excellent eyesight and seem to be studying you as you study them. They will build webs for shelter and child rearing but obtain food through stalking and ambush. Watch a jumping spider as it hunts and you will see it observe, plan, and attack carefully (although not always successfully.) Although the jumpers have formidable fangs and do have venom, you have nothing to fear from them. They will gladly spend time on your hand or sleeve, sometimes jumping from finger to finger before going on their way. Some of the jumpers are notable for their brightly colored chelicera ("jaws" that terminate in fangs) but they rarely bite humans, and the bites are not harmful. The following 16 species have all appeared in my yard.

Colonus Sylvanus
7 mm .

Bronze Jumping Spider
(Eris militaris)
7 mm.

Eris militaris in defensive posture.

Hentzia mitrata
4 mm.

Hentzia palmarum (M)
5 mm.

Hentzia palmarum (F)
6 mm.

Dimorphic Jumping Spider
(*Maevia inclemens*)
7 mm.

Emerald Jumping Spider
(*Paraphidippus aurantius)*
7 mm.

Peppered Jumping Spider
(Pelegrina galathea)
4 mm.

Brown and White Jumping Spider
(*Pelegrina proterva*)
4 mm.

Bold Jumping Spider
(Phidippus audax)
12 mm.
Our largest and most common jumper.

Brilliant Jumping Spider
(Phidippus clarus)
8 mm.

Phidippus insignarius (Male)
10 mm.

Canopy Jumping Spider
(Phidippus otiosus)
12 mm.

Phidippus princeps
9 mm.

Tan Jumping Spider
Platycryptus undatus
10 mm.

Zebra Jumping Spider
(*Salticus scenicus*)
5 mm.

Zygoballus sexpunctatus (male)
3-4 mm.

This is the smallest of our local jumpers, measuring about 3 mm. The chelicera has a unique inner tooth, giving it the appearance of having two sets of fangs.

Family Tetragnathidae – Long-jawed Orbweavers

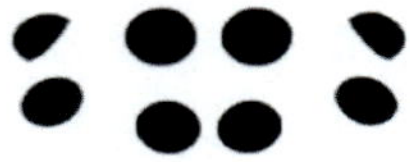

Tetragnatha Eye Pattern

Long-jawed Orbweavers have narrow bodies and, yes, long jaws. This long-jawed orbweaver is most often found near water, where it builds a horizontal web to catch insects or small arthropods. I have found many living in my kayak. This particular spider was under a deck, well away from water. This is probably T. versicolor, as that is the most commonly reported species in this area. However, definitive identification requires microscopic study.

Tetragnatha sp .
(Possibly *Tetragnatha versivolor*)
10 mm.

Family Theridiidae – Cobweb Spiders

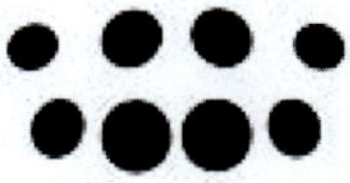

Theridiidae Eye Pattern

The Theridiidae family, also known as Tangle-web or Comb-footed Spiders, include the spiders most often found in human dwellings. The family includes medically important species such as the Black Widow (*Latrodectus mactans*) and is relatively well studied.

Enoplognatha caricis
(or *E. marmorata*)
5 mm.

Southern Black Widow
(*Latrodectus mactans*)
10 mm.

Caution: Do not handle! Will bite if threatened. Bite is not fatal but can be severe; should be seen by medical personnel.

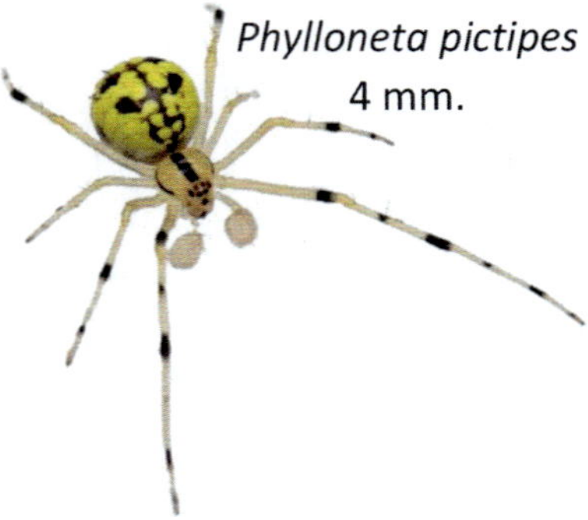

Phylloneta pictipes
4 mm.

Common House Spider
(*Parasteatoda tepidariorum*)
6 mm.

Steatoda borealis
5 mm.

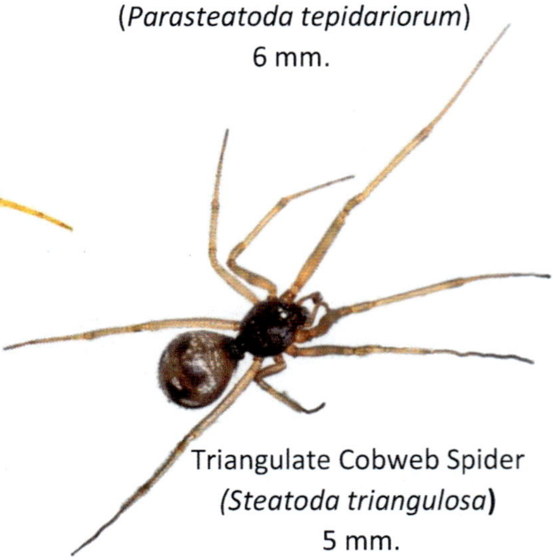

Triangulate Cobweb Spider
(Steatoda triangulosa)
5 mm.

Theridula opulenta
2 mm.
Quite small but distinctive, this individual was found in its web under a leaf about 6' above ground.

Theridiidae sp. with egg sac

Family Thomisidae – Crab Spiders

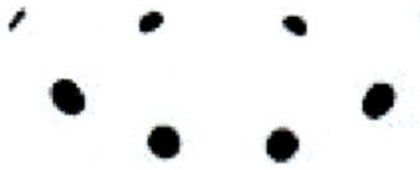

Thomisidae Eye Pattern

Distinguished by their crab-like appearance, there are 2100 species worldwide. In addition, some spiders that appear to have a similar shape are locally called "crab spiders." Crab Spiders are sometimes called flower spiders or flower crab spiders. These spiders don't bother with webs (although they do produce silk that they may use for other purposes.) Some actively hunt for food but most, like this *Mecaphesa* sp., are ambush predators.

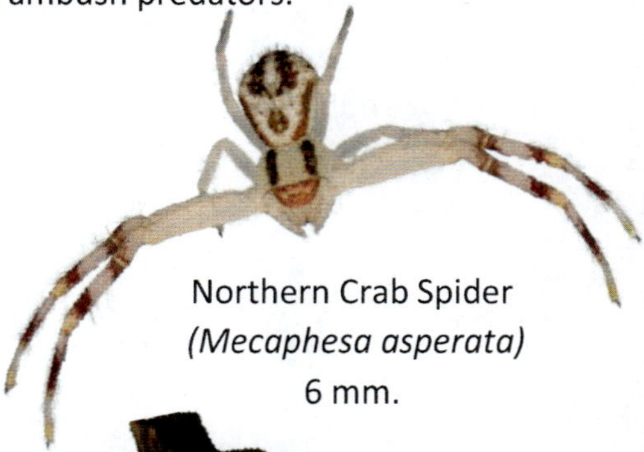

Northern Crab Spider
(Mecaphesa asperata)
6 mm.

Mecaphesa sp

(**F** 5-11 mm.) White-banded Crab Spider (M 2-3 mm.)
(*Misumenoides formosipes***)**

Synema parvulum
4 mm.

*Thomisidae sp***.**
7 mm.

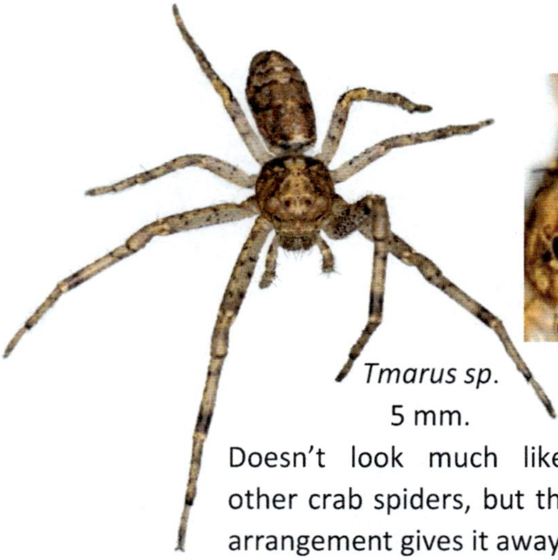

Tmarus sp.
5 mm.
Doesn't look much like the other crab spiders, but the eye arrangement gives it away.

Xysticus sp.
6 mm.

Family Philodromidae – Running Crab Spiders

Philodromidae Eye Pattern

Philodromus sp .

5 mm.

Possibly *P. cespitum* which is common in European orchards, occasionally found in the United States.

Sub-class Acari – Ticks and Mites

50,000 described species; one million more est.

Order Trombidiforms - Mites

Family Erythraeidae

Larval Erythraeidae are parasites of spiders and specific insects, including honeybees.

Callidosomatinae sp.
1 mm.
A long-legged velvet mite

Family Trombidiidae – True Velvet Mites

Trombidioidea sp.
1 mm.
A True Velvet Mite
Aka Red Velvet Mite or Rain Bug

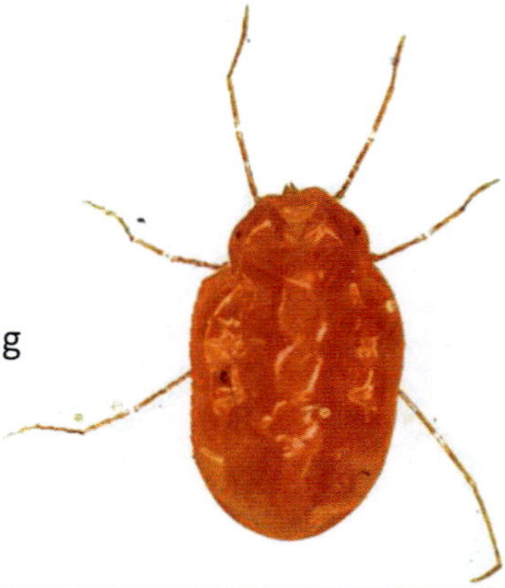

Hydrachnidia, an Infraorder containing Water Mites
1 mm.

Order Ixodida – Ticks

Family Ixodidae – Hard Ticks

Three species are common in our area: Lone Star tick, American dog tick, and the small deer tick (aka black-legged tick.) Larval ticks do not carry diseases, but after feeding on infected hosts they can transmit a variety of nasty diseases. In some cases, transmission of disease takes many hours or days, so timely removal of ticks is important.

Lone Star Tick
(*Amblyomma americanum)* nymph
0.5 mm.

Engorged females can be about 15 mm. Adults and nymphs may transmit Monocytic Ehrlichiosis, Rocky Mountain Spotted Fever and 'Stari' borreliosis.

American Dog Tick
(*Dermacentor viriabilis)* – male & female

Adults and, nymphs may transmit Rocky Mountain spotted fever and tularemia.

Order Opiliones – Harvestmen

Harvestmen are arachnids but not spiders and contrary to folk tales they do not possess lethal venom. Their fangs are so small they couldn't inject anything into humans in any case. Known by various names (e.g., daddy long-legs), they come in many shapes and sizes and go quietly about their business eating smaller bugs and forest detritus.

Leiobunum sp.

Order Pseudoscorpione – Pseudoscorpions

Looking somewhat like very small scorpions (2-4 mm.), the Pseudoscorpions are beneficial predators of book lice, mites, and assorted mini-pests. By attaching themselves to larger insects, they disperse while using little energy, an interaction called "phoresy". More than 3300 species are found throughout the world. These individuals were found in leaf litter.

Apochthonius sp.

Unidentified species.

Class Insecta – Insects

Insects are six-legged invertebrates and the largest group within the arthropod phylum. Insects have a chitinous (pronounced "**Ky**-ten-us") exoskeleton, a three-part body (head, thorax, and abdomen), three pairs of legs, compound eyes, and one pair of antennae. There are about a million described species, and perhaps six to ten million undescribed or undiscovered. Most insects hatch from eggs and many undergo a four-stage development (holometabolism) consisting of egg-larva-pupa-adult. Others undergo a three-stage development (hemimetabolism) of egg-nymph-adult. In holometabolous insects the various nymphal stages (instars) may look vastly different from each other and have little resemblance to the adult. Hemimetabolous nymphs look much like adults but lack wings and reproductive structures.

Order Blattodea – Cockroaches and Termites

Family Ectobiidea – Cockroach

Most described species of cockroach live in tropical climates; 40 are known in the U.S. Of those, the German cockroach (Blattella germanica) is the most familiar and reviled for its habit of moving into our kitchens and pantries. The vast majority of cockroaches are beneficial composters and stay out of our sight and out of our homes.

Wood Cockroachs
(Parcpblatta sp)

Family Rhinotermitidae – Subterranean Termites

Nine species of Rhinotermitidae are found in the U.S. and some are likely found in your yard. In nature they are important recyclers; in your walls they are never welcome. Keeping moisture away from the foundation of the house will help deter termites, but termites swarm and can find their way into almost any habitat that is dark, moist, and contains cellulose (wood, cardboard, etc.)

Factoid: Termites evolved from cockroaches during the Jurassic period.

Eastern Subterranean Termite
(*Reticulitermes flavipes*)
8 mm.

Order Coleoptera (And the Beetles go on)

With beetles we've hit the jackpot! There are about 400,000 species of beetles described and no doubt many more waiting in the shadows. Beetles constitute 40% of all described insect species. For each mammal species there are 100 beetle species. Chances are there are many different beetles in your yard, most of which you have never noticed.

Family Aderidae – Ant-like Leaf Beetles

As the name suggests, these small beetles vaguely resemble ants; often found on the underside of leaves.

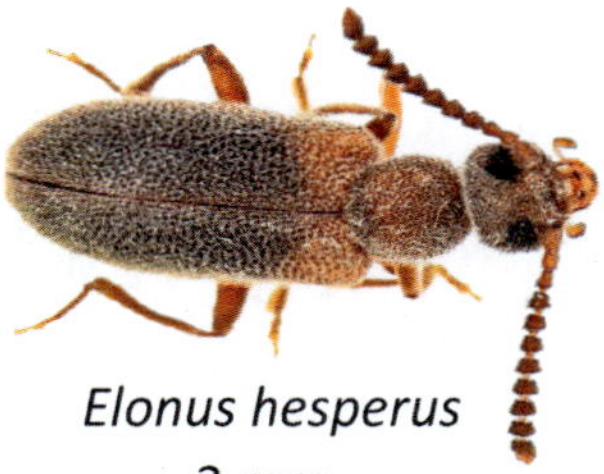

Elonus hesperus
3 mm.

Family Anthicidae – Ant-like Flower Beetles

Cloudy Flower Beetle
(Anthicus cervinus)
3 mm.

Family Anthribidae – Fungus Weevils

Small weevils with straight, occasionally long antennae. Diet includes fungi and decaying plant material as adults, dead wood as immature.

Euparius sp (E. paganus?)

Family Brentidae – Primitive Weevils

Oak Timberworm
(Arrenodes minutus)
16 mm.

The Oak Timberworm (*Arrenodes minutus*) is the only Primitive Weevil found north of Florida. The male has a short, broad snout; the female has a longer, slender snout. As the name suggests, the preferred host for this weevil is oak where it drills into the tree to lay eggs.

Family Buprestidae – Metallic Wood-boring Beetles (aka Jewel beetles)

15,000+ species, 78 reported in the U.S. They are, as the name suggests, a shiny and colorful group. Adult *buprestis* lives in dead or dying hardwoods, specializing in maple, beech, blackgum, oak and elm. Larval *buprestis* burrow under the bark of trees and may live there for years before becoming adults. Normal populations of *buprestis* will not substantially weaken the tree, but if the tree is being grown to be harvested for lumber, the burrows made by larvae will devalue the wood. My informal survey of the neighborhood found nobody growing oak trees for harvest.

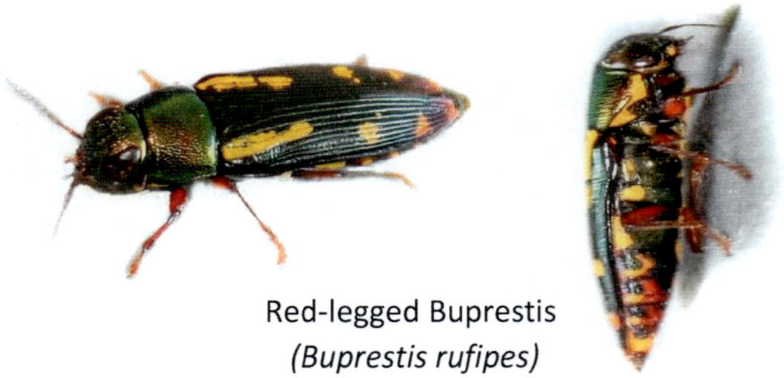

Red-legged Buprestis
(Buprestis rufipes)
20 mm.

Dicerca obscura
13-20 mm.
Larvae found in common persimmon and sumac; adults on hickory, oak.

Family Cantharidae – Soldier Beetles

Aka "Leatherwings", soldier beetles are small, slender, and numerous (more than 500 species in the U.S.)

Two-lined Leather-wing
(Atalantycha bilineata)
8 mm.
aka Two-lined Cantharid

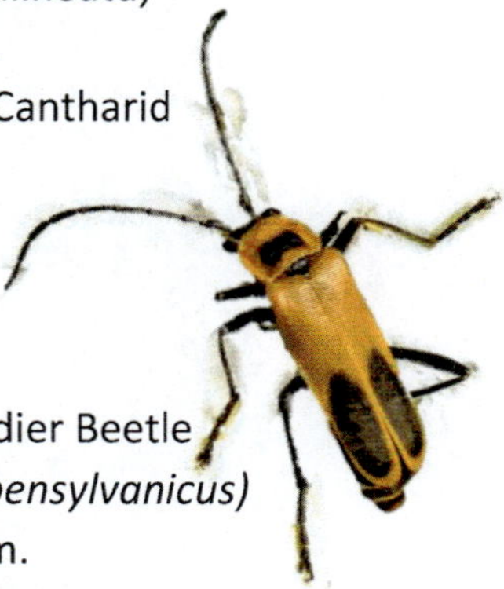

Goldenrod Soldier Beetle
(Chauliognathus pensylvanicus)
16 mm.
Quite common

Margined Leatherwing
(*Chauliognathus marginatus*)
10 mm.

Wrinkled Soldier Beetle
(*Podabrus rugosulus*)
7-9 mm.

Family Carabidae – Ground Beetles

Most of the 2,000+ species of Ground Beetles in the U.S. are shiny black or metallic with ridged wing-covers (elytra). Most are capable of producing defensive foul-smelling acidic secretions; some species are capable of shooting these chemicals when threatened. Generally considered beneficial, as they prey of the larvae of other garden insects.

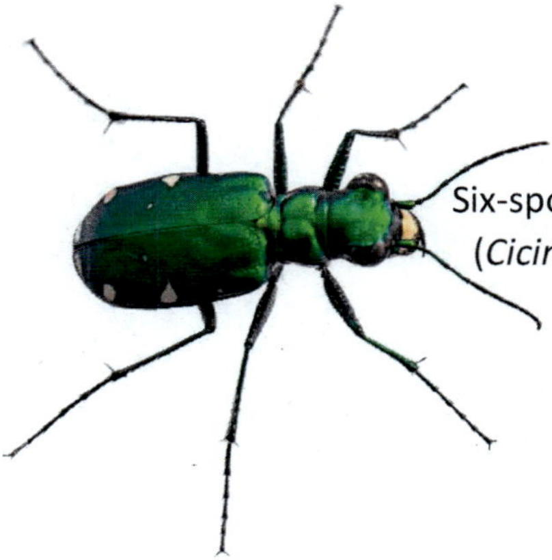

Six-spotted Tiger Beetle
(*Cicindela sexguttata*)
10-14 mm.

Virginia Bigheaded Tiger Beetle
(*Tetracha virginica*)
16-25 mm.

Punctured Tiger Beetle
(*Cicindelidia punctulata*)
10-13 mm.

Chlaenius aestivus
14-17 mm.
A Vivid Metallic Ground Beetle

Chlaenius laticollis
14 mm.

Chlaenius sp. nymph

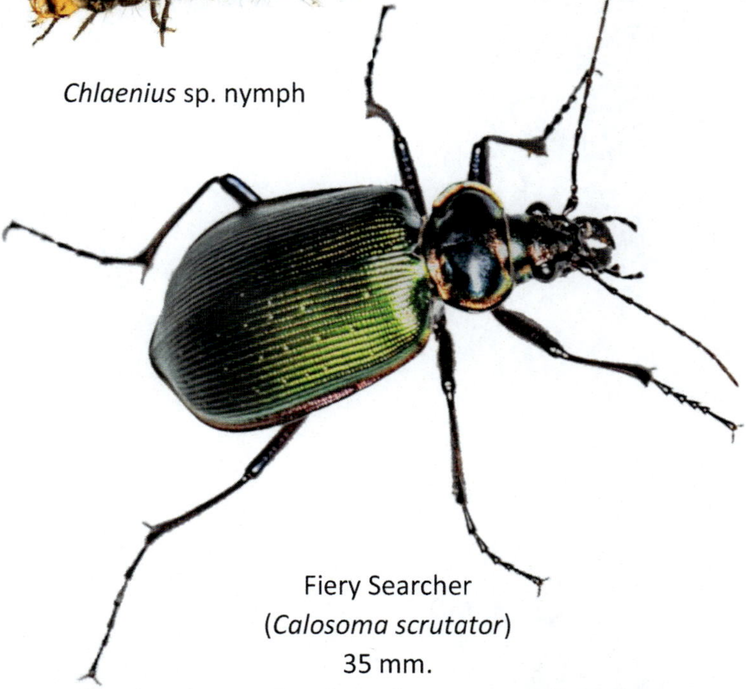

Fiery Searcher
(*Calosoma scrutator*)
35 mm.

Impressive in size and color, *C. scrutator* is uncommon and unforgettable. Observed only once in 2020. Also called “Caterpillar hunter,” *C. scrutator* has an appetite for tent caterpillars and gypsy moths but will eat a variety of caterpillars and larvae. It is a top predator of the insect world.

Cosnania pensylvanica
7 mm. ant mimic

Dromius piceus
7 mm.

Harpalus sp.
23 mm.

Lebia analis
5 mm.

Lebia grandis
10 mm.

Lebia viridis
6 mm.

Poecillus chalcites
10 mm.
A woodland ground beetle

Pterostichus sp.
More than 1,000 species in the genus. Most are predatory, though some feast on strawberries.

Pedunculate Ground Beetle
(*Clivina sp.*)
10 mm.

Selenophorus ochropezus
5 mm.

LeConte's Seedcorn Beetle
(Stenolophus lecontei)
6 mm.

Family Cerambycidae – Long-horn Beetles

This large group of beetles is well represented in our area. Most longhorn beetles have quite long antennae, often longer than their bodies. Most *cerambicidae* larvae feed on woody tissue, while their adults are nectar eaters.

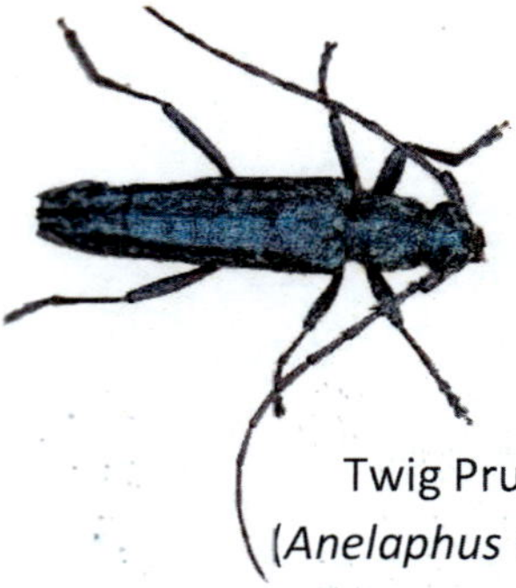

Astyleiopus variegatus
7-11 mm.

Twig Pruner
(*Anelaphus villosus*)
14 mm.

Spined Bark Borer
(*Elaphidion mucronatum*)
a.k.a. Spined oak borer
25 mm.

Hippopsis lemniscate
10 mm.

Enaphalodes atomarius
an oak borer
22 mm.

Red-headed Ash Borer
(Neoclytus acuminatus)

Locust Borer
(*Megacyllene robiniae*)
20 mm.
Larvae feed on black locust, tunneling under the bark.

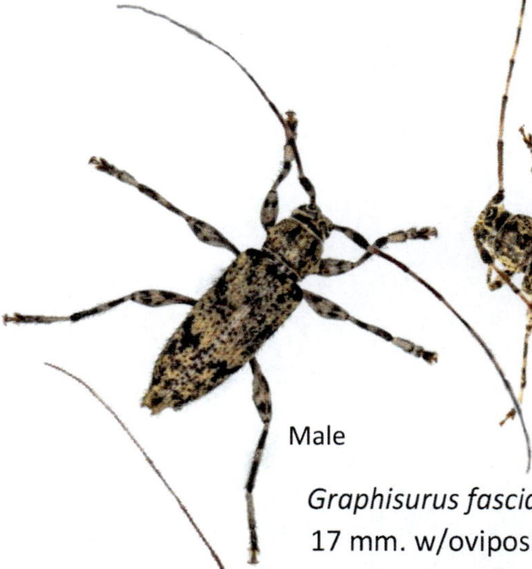

Male

Female

Graphisurus fasciatus
17 mm. w/ovipositor
Feeds on hardwoods and pine

Pine Sawyer Longhorn Beetle
(Monochamus carolinensis)

Tile-horned Prionus male
(Prionus Imbricornis)
40 mm.
Larvae feed in roots (oak, chestnut)

Slender Flower Longhorn Beetle
(*Strangalia famelica*)
9 mm.

Stenocorus cinnamopterus
9 mm.

Ash and Privet Borer
(Tylonotus bimaculatus)
9 mm.

Red Milkweed Beetle
(*Tetraopes tetrophthalmus)*
12 mm.

Family Chrysomelidae – Leaf Beetles

Generally small, often quite colorful, the leaf beetles munch on herbaceous plants. Several agricultural pests reside in this family.

Striped Cucumber Beetle
(*Acalymma vittatum*)
5 mm.

A major pest of back yard gardens, the adults will eat leaves, flowers and rinds of cucumbers and related crops; the larva will tunnel into the stems and vines. These will show up in large numbers even if you are planting your first cucumbers and there are no other plants in the neighborhood!

Acanthoscelides obtectus – a Bean Weevil
4 mm

Native to South America but now widespread, *A. obtectus* can be a major pest of beans and related legumes in the field and in storage. Yes, a bean weevil is a Chrysomelid, not a Curculionid like the rest of the weevils.

Basswood Leaf Miner
(*Baliosus nervosus*)
6 mm.

Normally feeds on the leaves of the American basswood tree, but known to mine the leaves of willows, poplars, and related species.

Brachypnoea sp
3 mm.
Found on a 'shroom

Charlie Brown Flea Beetle
(*Capraita sexmaculata)*
3 mm.

The Mottled Tortoise Beetle (*Deloyala guttata*) nymph caries a fecal shield for protection. The adult sports a translucent shield of its own.

6 mm.

Bean Leaf Beetle
(*Cerotoma trifurcate*)
5 mm.

Dogbane Leaf Beetle
(*Chrysochus aurantus*)
8-13 mm.

This all-green beetle reflects light in wonderful ways.

A Flea Beetle
(Crepidodera sp)
3 mm.
Prob. *Crepidodera nana* or C. *luminosa*.)

Grape Colaspis
(*Colaspis brunnea*)
4 mm.

Demotina modesta
2 mm.
Definitely a female. D. modesta is parthenogenetic—there have been no males found in the US. Often found on oak.

Cucumber Leaf Beetle
5 mm.
(*Diabrotica undecimpunctata)*

Aka Spotted Cucumber Beetle or Southern Corn Rootworm, this bug can cause considerable damage to your crops. The larvae eat the roots of young plants, while the adults eat the flowers, foliage, and fruit.

Swamp Milkweed Leaf Beetle
(*Labidomera clivicollis*)
12 mm.
A large, conspicuous beetle specializing in milkweed.

Rhabdopterus praetextus
5 mm.
Note: Species identification requires inspection of the male genitalia; females cannot be identified except by association with the males. None of that was done. This bug is likely R. praetextus based on known distribution of this species.

Cryptocephalus badius
4 mm.
A case-bearing leaf beetle

Neochlamisus bebbianae – a Warty Leaf Beetle
4 mm.

Waterlily Leaf Beetle
(*Galerucella nymphaeae*)
5 mm.
Many of these found on water lily leaf pads. Eggs are laid on top of the lily pads in clusters.

Family Cleridae – Checkered Beetles

Narrow and hairy, checkered beetles fill many ecological niches. Some are used as biological controls, as they have a healthy appetite for bark beetles: adults eating adults, larvae eating larvae.

Priocera castanea
6 mm.

Madoniella dislocata
3-6 mm.

Family Coccinellidae – Lady Beetles

Known as ladybugs in the U.S. and ladybird beetles in the rest of the English-speaking world, lady beetles are distinctive, beneficial, and beloved. Most are consumers of aphids and scale insects, although some are predators of other lady beetle nymphs.

Seven-spotted Lady Beetle
(*Coccinella septempunctata*)
7 mm.

Spotted Lady Beetle
(*Coleimegilla maculata lengi*)
5 mm.

Polished Lady Beetle
(*Cycloneda munda)*
4 mm.

Asian Lady Beetle
(*Harmonia axyridis*)
5-8 mm

Imported for pest management, *H. axyridis* is larger and more aggressive than most native lady beetles. The larvae and adults eat the larvae of native lady beetles as well as aphids. As a result, *H. axyridis* is now the dominant species in most of their range. This species is quite variable in pattern.

Twenty-spotted Lady Beetle larva
(Psyllobora vigintimaculata)

Scymnus sp

Not all lady beetles are red. Some are dark and hairy.

Families Cucujidae and Silvanidae – Flat Bark Beetles

Family Cucujudae

Red Bark Beetle
(Cucujus clavipes)
12 mm.

Family Silvanidae

Telephanus atricapillus
4 mm.

Family Curculionidae – Snout[9], Ambrosia, and Bark Beetles

Euwallacea validus

Euwallacea Validus
2 mm.

An invasive ambrosia beetle from Japan, first noted in New York in 1975, is now broadly established in the United States. These beetles bore into the phloem layer of various trees, where they release spores and cultivate fungus gardens for food. This one was found on an old, weakened cherry tree. Looking unlike the snout beetles (weevils), ambrosia beetles were formerly classified as their own family, Platypodidae. They are now considered Scolytinae or Platypodinae, sub-families of Curculionidae.

Pecan Gall Curculio
(*Conotrachelus elegans*)
4 mm.

[9] Weevils

Cylindrocopturus quercus
2 mm.

Corn Pith Weevil
(Geraeus penicillus)
2 mm.

Listronotus sp.
looks like *L .appendiculatus*
6 mm
This small weevil had even smaller beetles (*Toramus pulchellus*) riding on its back.

Rhubarb Weevil
(*Lixus concavus*)
12 mm

As you would guess, the Rhubarb weevil is known to cause damage to rhubarb by cutting into the stems although it is rare to find this species producing economically significant damage to crops. The larva may damage roots of crops.

Micralcinus sp.
M. parvulus or *M. stehi*
3 mm.

White-fringed Weevil
(*Naupactus leucoloma)*
10 mm.

Yellow Poplar Weevil
(*Odontopus calceatus)*
4 mm

A.k.a. sassafras mining weevil, tulip tree leaf miner, tulip tree weevil, or the magnolia leaf miner. Adults emerge in the spring after overwintering in the ground. They lay eggs on the leaves of host plants (yellow poplar, magnolia...). By mid-summer the eggs have hatched, the larvae have pupated, and the adults disappear into the leaf litter till next year. Although not usually harmful to their host, occasional infestations have caused significant damage to yellow poplar.

Pseudocneorhinus bifasciatus, the Two-banded Japanese Weevil, was introduced from Asia about 100 years ago and is now well established throughout the eastern U.S. With more than 100 host plants, this weevil seems to have a particular fondness for roses, hemlock and laurel. All *P. bifasciatus* in the U.S. are female, reproducing parthenogenetically.

Two-banded Japanese Weevil
(Pseudocneorhinus bifasciatus)
3 mm.

Another Asian import, *Pseudoedophrys hilleri* tends to prefer maple and witch hazel. That is certainly the case of these particular individuals who are abundant under a maple tree, although they have not caused noticable damage to the tree.

Pseudoedophrys hilleri
4 mm.

Ironweed Curculio
(*Rhodobaenus tredecimpunctatus)*
10 mm.

Sphenophorus melanocephalus
A 10 mm. billbug
Grubs will eat the roots of turf grass. The grubs look like small Japanese beetle grubs but are straight rather than "C"-shaped; they have no legs.

Maize Weevil
(*Sitophilus zeamais)*
3 mm.

Fourteen species of the Sitophilus (granary weevil) genus are found throughout the world feasting on grains and nuts, causing considerable loss of stored grains. This individual showed up in a sack of feed corn, so is probably *Sitophilus zeamais*, the maize weevil.

Xyleborus xylographus
3 mm.
A bark beetle. Looking unlike the snout beetles (weevils), bark beetles were formerly classified as their own family, *Scolytidae*. They are now considered *Scolytinae*, a sub-family within Curculionidae.

Family Dermestidae – Skin Beetles

Although adult skin beetles feed on plants, the larvae feed on dried skin, fur, fibers, and feathers. Some species are useful in forensic analysis.

Anthrenus sp
3 mm.

Anthrenus verbasci
3 mm.

Trogoderma sp.
a carpet beetle
3 mm.

Family Elateridae – Click Beetles

A large family whose shared characteristic is the ability to snap their bodies with a "click", propelling themselves into the air. While the adults have little agricultural impact, several species' larvae (wireworms) cause significant damage.

Flat Wireworm
(*Aeolus mellilus)*
6 mm.

Eyed Click Beetle
(*Alaus oculatus*)
40 mm.
The largest of our click beetles

Conoderus bellus
5 mm.

Conoderus lividus
11-17 mm.

Elater abruptus
15-22 mm.

Gambrinus griseus
10-14 mm.

Melanotus sp
5-12 mm.

Orthostethus infuscatus
30 mm.

Family Elmidae – Riffle Beetles

Larvae and adults are generally aquatic, living under rocks and eating algae.

Stenelmis sp.

7 mm.

Although most riffle beetles are aquatic, this individual found its way ashore (about 25’ from the lake.)

Family Erotylidae – Pleasing Fungus Beetles

A beetle of little impact on human endeavors, this family goes about life eating fungi and plant material; some are pollinators.

Four-spotted Fungus Beetle
(Ischyrus *quadripunctatus)*
9 mm.

Megalodacne fasciata
9-15 mm.

Toramus pulchellus
1 mm.

Family Mycetophagidae – Hairy Fungus Beetles

Another family of fungus eaters, the Myvetphagidae live quietly among the leaf litter and fungi.

Litargus tetraspilotus
2 mm.

Family Gyrinidae – Whirligig Beetles

The most visible of the aquatic beetles, the whirligigs are easily seen in groups swimming on the surface of the water. When disturbed they swim in a frenzied circular pattern, hence the "whirligig." The eyes are split, allowing sight simultaneously above and below water.

Dineutus sp
11 mm.

Family Haliplidae – Crawling Water Beetles

This family of water beetles does not swim well, preferring to crawl. Their swimming is accomplished by paddling alternately, creating a clumsy-appearing motion. This may be beneficial as a means of confusing predators.

Peltodytes muticus
5 mm.

Family Heteroceridae – Variegated Mud-loving Beetles

Small beetles that dig (sometimes tunnel) through mud at the edge of ponds, eating algae.

Heterocerus mollinus
5 mm.

Family Histeridae – Clown Beetles

Predacious beetles with short elytra (wing covers), clown beetles will eat larvae, munch on carcasses, and eat each other.

Aeletes politus
1 mm.

Euspilotus assimilis
5 mm.
A common carrion-feeder

Family Hybosoridae – Scavenger Scarab Beetles

A Pill Scarab Beetle
(*Germarostes globosus)*
3 mm.
Nearly identical to *Germarostes aphodioides*

Family Hydrophilidae – Water Scavenger Beetles

The water scavenger beetles are misnamed, as the larva are predators and, while some adults will scavenge, most will predate as well. Mosquito larvae are a favorite of some *Hydrophilidae*.

Cercyon praetextatus
8 mm.

Collared Water Scavenger Beetle
(*Tropisternus collaris)*
8 mm.

Berosus pantherinus
4 mm.

Family Lucanidae – Stag Beetles

Large, distinctive beetles. The males have oversized mandibles reminiscent of the antlers of a stag. Larvae of the stag beetles may live several years and several larval stages before pupating. The males may look intimidating, but they are not aggressive toward humans. The smaller females have much smaller (and more powerful) mandibles.

Yellow-thighed Stag Beetle male
(*Lucanus capreolus*)
a.k.a. Reddish-brown Stag Beetle
28 mm.

Family Meloidae – Blister Beetles

Adult blister beetles are all capable of producing cantharidin, a blister agent, for defense. These insects may inhabit grasses—especially alfalfa- - which could be processed into hay bales. A few beetles consumed in a single feeding may be lethal to a horse.

Margined Blister Beetle
(*Epocauta funebris*)
10 mm.
Pest of tomatoes or potatoes.

Black Blister Beetle
(*Epocauta pensylvanica*)
11 mm.

Striped Blister Beetle
(*Epocauta vittata*)
9 mm.

The larvae remain mostly underground, feeding on grasshopper eggs. The adults, however, are a major pest of squash, beans, and numerous other crops. They tend to appear in swarms. If you pick these off your plants, wear gloves. They are blister beetles, after all.

Family Melyridae – Soft-winged Flower Beetles

Primarily pollen-eaters as adults, the Melyridae are sometimes called "pollen beetles."

Attalus sp
6 mm.

Family Mordellidae – Tumbling Flower Beetles

To escape predators the tumbling flower beetle will initiate a series of seemingly uncoordinated jumps, appearing to tumble. Due to the long abdominal tip that facilitates these movements, they are sometimes called pintail beetles.

Falsomordellistena pubescena
4 mm.

Family Nitidulidae – Sap-feeding Beetles

Small beetles that feed on sap or over-ripe fruit.

Picnic Beetle
(*Glischrochilus fasciatus*)
5 mm.

Attracted to beer, wine, fruity or fermenting beverages, the picnic beetle earns its name by drinking from and drowning in your beer.

Glischrochilus sanguinolentus
5 mm.

Glischrochilus siepmanni
6 mm.

Four-spotted Sap Beetle
(*Glischrochilus quadrisignatus*)
5 mm
Associated with milkweed seed pods, this insect is sometimes referred to as "beer bug" or "picnic beetle for its habit of finding and drowning in your beer.

Strawberry Sap Beetle
(*Stelidota geminata*)
3 mm.
Quickly attracted to damaged or over-ripe strawberries, *S. geminate* can become a pest in the garden.

Family Ptilodactylidae – Ptilodactylid Beetles

Philidactyl sp,
5 mm.
A consumer of fungi and decaying leaf litter.

Family Ptinidae – Death Watch & Spider Beetles

Caenocara sp, a puff-ball beetle
2 mm.
Small beetle that develops within puffball fungi.

Family Pyrochroidae – Fire-colored Beetles

Also known as "eyelash bugs," these are primarily fungus-eaters

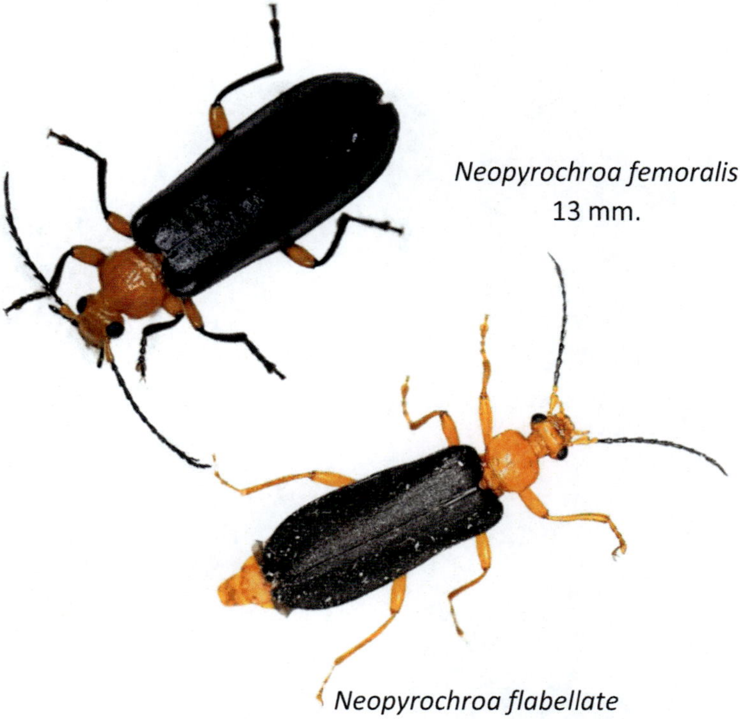

Neopyrochroa femoralis
13 mm.

Neopyrochroa flabellate
13 mm.
A nocturnal fungivore, spends days under decaying logs.

Family Scarabaeidae – Scarab Beetles

A large and important family, scarabs fill many niches. Some are recyclers of dung (dung beetles) while others are plant eaters (chafers, Japanese beetles.) The scarab revered in ancient Egypt is a dung beetle.

Black Turfgrass Ataenius
(Ataenius spretulus)

Green June Beetle
(*Cotinis nitida***)**

Northern Masked Chafer
(Cyclocephala borealis)

Dark Flower Scarab
(*Euphoria sepulcralis*)

Oriental beetle
(*Exomala orientalis*)
7-13 mm.

Eastern Hercules Beetle (F)
(*Dynastes tityus*)
52 mm.

Rose Chafer
(*Macrodactylus subspinosus*)
7-10 mm.

Asiatic Garden Beetle
(*Maladera formosae*)
12 mm.

Scooped Scarab Beetle
(*Onthophagus hecate*)
A dung beetle

Grapevine Beetle
(*Pelidnota punctata*)
24 mm.

Japanese Beetle
(*Popillia japonica*)
11 mm.

Few beetles are as common, numerous, and destructive as the Japanese beetle. The adult has an appetite for more than 300 plants, and its grubs are a major destroyer of turf.

Family Silphidae – Carrion Beetles

Where would we be without carrion beetles? Knee-deep in carcasses? Probably not. There are a host of other recyclers at work. However, carrion beetles are one of the first, fastest, and most efficient recyclers of animal bodies. This family is also noted for parenting behavior.

American Carrion Beetle
(*Necrophila americana*)
20 mm.

nymph

While flies are often the first insects at a corpse, *N. americana* arrives soon after and provides an important service: Not only does it eat the carrion, it also eats the fly larvae. The beetle lays and guards its eggs; the nymphs are active after other recyclers have departed, so consume more dried hide, hair, or feathers. One generation per year.

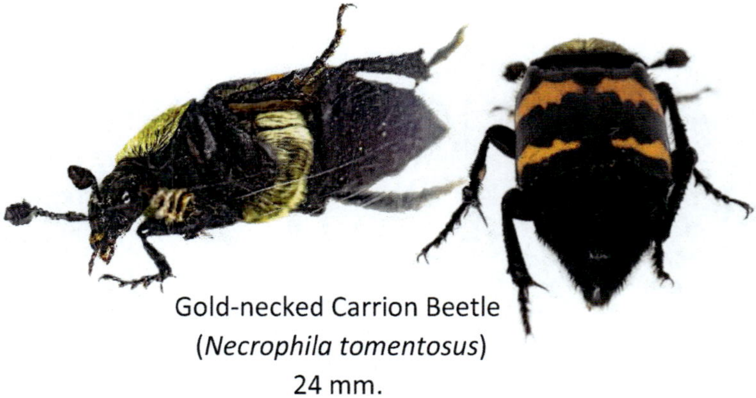

Gold-necked Carrion Beetle
(*Necrophila tomentosus*)
24 mm.

Ridged Carrion Beetle
(*Oiceoptoma inaequale*)
14 mm.

Margined Carrion Beetle
(*Oiceoptoma noveboracense*)
12 mm.

Family Staphylinidae – Rove Beetles

With 63,000 members, the rove beetles comprise the largest family of organisms. They are predators, generally long and thin with short elytra.

Aleochara littoralis (?)
A parasitic rove beetle
6 mm.

Apocellus sphaericollis
3 mm.
A spiny-legged rove beetle
The only *Apocellus* species in North America

Astenus sp.

Belonuchus rufipennis
7 mm.

Bisnius blandus
3 mm.

Coproporus ventriculus
3 mm.
a crab-like rove beetle

Falagria dissecta
2 mm.

Homaeotarsus sp.

Paederus littorarius
6 mm.
Caution: Do not crush! *Paederus* sp. contain pederin which will cause severe blistering if it gets on your skin.

Scopaeus sp

Lathrobiina (sub-tribe) sp

Pinophilus sp.
probably *Pinophilus latipes*
14 mm.

Family Tenebrionidae – Darkling Beetles

A large family of dark beetles, primarily scavengers of dead wood and debris but also know to dine on other insects. Larvae resemble click beetle larvae.

False Mealworm Beetle
(*Alobates pensylvanicus*)
21 mm.

Family Trogidae – Keratin (Hide) Beetles

Among the scavengers, Hide beetles are late arrivals at any carcass. They specialize in hide, hair, and feathers. Some may be found in bird nests. These bugs are not widely studied.

Trox unistriatus
11 mm.
Found on a fox carcass

Order Dermaptera – Earwigs

Family Forficuladae

European Earwig
(Forficula Auricularia) -- Immature male
12 mm

Introduced to North America in the early 1900s, *F. auricularia* is now the dominant species in many areas. The mature male will have "spurs" near the base of its pincers. Females have straight pincers.

Order Diptera – Flies

The estimated one million species of diptera have been so named because they have a single pair of wings. The second set of wings found on most other flying insects has evolved into a stabilizing structure ("haltere") in the flies. 125,000 species have been described, including crane flies, horseflies, mosquitos, houseflies and more. They are a nuisance in your house and at your picnic...some are a major vector of disease...but they also play an important role in the environment as recyclers, pollinators, predators, and, of course, food.

Family Asilidae – Robber fly

Stout bodies, spiny legs and piercing mouth parts make the robber fly and effective predator of most any insect of appropriate size. They may even take prey larger than themselves.

Atomosia sp
6 mm.

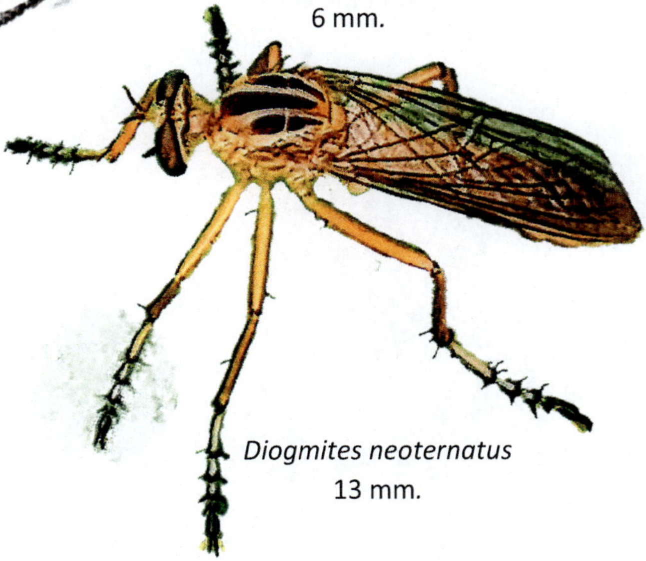

Diogmites neoternatus
13 mm.

Efferia aestuans (M)
16 mm.

Holcocephala sp
3 mm.

Locally called "gnat ogres," these small robber flies are well represented in tall grass or wingstem patches on the edges of fields.

Robber Fly (F) with prey

Taracticus octopunctatus -- 7 mm.

Family Bibionidae – March Flies

Bibio articulatus – 5 mm.

Tiny flowers require tiny pollinators. Here's *Bibio articulatus*, a March fly female measuring about 5 mm (1/5-inch), perfectly suited for the job. March flies are short-lived as adults, generally living less than a month, and are seen (or unseen, as they are so small) through early summer. Their larvae feed on plant roots and detritus, conditioning the soil. This is also an example of why binomial nomenclature is important. *Bibio articulatus* is the same bug anywhere you find her. In Australia, however, if you talk about March flies, you're talking about Tabanids. We call those things horseflies!

Family Bombtliidae – Bee Flies

Bee fly adults are pollen eaters and pollinators. Their larvae are often parasites of beetles or flies.

Tiger Bee Fly (*Xenox tigrinus*)
16 mm.

Tiger bees lay eggs near the entrance of carpenter bee nests. The larvae enter the nest and feed on carpenter bee eggs and pupae.

*Bombylius s***p**

Family Ceratopogonidae – Biting Midges

Palpomyia sp
5 mm.

Family Chaoboridae – Phantom Midges

Chaoborus punctipennis - ♂
3 mm.

Family Chironomidae – Midges

Often mistaken for mosquitoes, *Chironomidae* do not bite. Their primary function in life seems to be food for just about any fish, fowl, spider, or predatory insect.

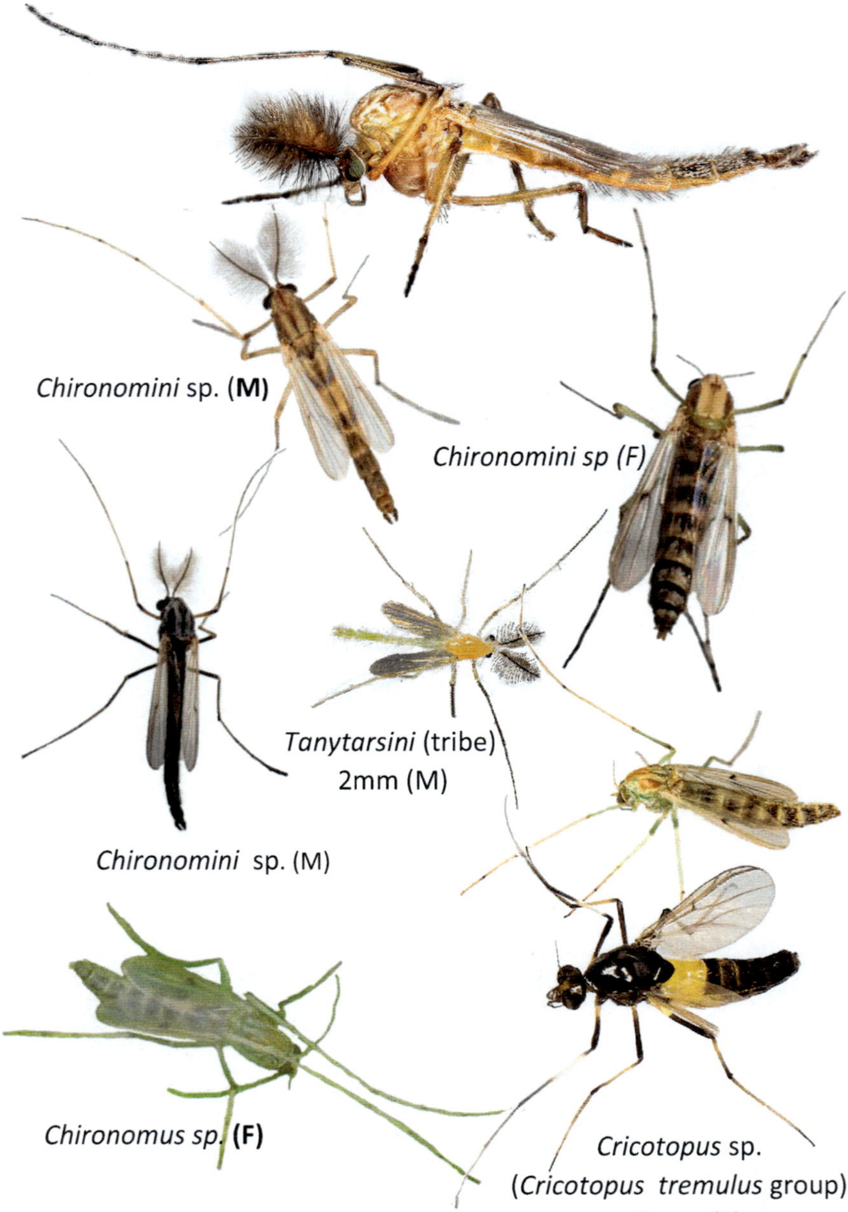

Chironomini sp. (**M**)

Chironomini sp (F)

Tanytarsini (tribe)
2mm (M)

Chironomini sp. (M)

Chironomus sp. **(F)**

Cricotopus sp.
(*Cricotopus tremulus* group)
5 mm (F)

Family Chloropidae – Frit Flies (a.k.a. Grass flies)

Small, hairless flies whose larva generally prefer grasses. Occasionally pests of cereal grains. Adults sometimes attracted to eye secretions.

Frit fly
3 mm.

Family Dolichopodidae – Long-legged Flies

Small, shiny, long-legged predators

Condylostylus sp.
(*caudatus* group)
6 mm.

Dolichopus sp
7 mm.

Family Drosophilidae – Fruit Flies

Fruit flies are rarely considered a pest but are often a nuisance. They are attracted to fermenting fruit (also beer & wine) in large numbers.

African fig fly
(*Zaprionis indianus*)
2 mm.

A native of Asia and Africa, *Z. indianus* has recently established itself in the Americas, working its way north. This individual was on an orange being offered to orioles.

Family Empidinae – Dance Flies

150 described *Rhamphomyia* species in North America (with an estimated 400 undescribed species), dance flies tend to swarm along the edges of freshwater ponds. Also known as Balloon Flies, Empids, and Dagger Flies, they use their considerable proboscis to subdue small insects.

Long-tailed Dance Fly
(*Rhamphomyia longicauda)*
6 mm.

Rhamphomyia sp
6 mm.

Family Fanniidae

Small flies, formerly in the *Musca* family with house flies

Fannia larva on bird carcass

Family Lauxaniidae

Lauxaniidae look a lot like Drosophilidae: The easiest distinguishing character to look for is the strong bristles on the top of the head right above the eye. Lauxaniids will have two strong bristles above each eye, both facing backwards. Drosophilids will have one strong bristle facing backward, and usually a strong one facing forward. However, these are small flies and the chances are slim they'll wait around while you count their eye lashes.

Lauxaniidie sp

Family Micropezidae – Stilt-legged Flies

Small, slender flies. Front legs often reduced, may be lightly colored on the ends. Some species wave the front legs above their heads to mimic wasps.

Rainieria antennaepes
10 mm.
This individual was observed eating fox scat.

Family Muscidae – House Flies & Kin

Face Fly (M)
(Musca autumnalis)
7 mm.

Closely related to the house fly (*Musca domestica*), the face fly is a nuisance of cattle, seeking secretion from the eyes and blood from horsefly bites. Introduced from Europe in the 1940s, widespread today.

Family Phoridae -- Scuttle Flies

Phoridae are the smallest of the flies, ranging from 0.5-6 mm. They are called "scuttle flies" for their habit of running at times when other flies would take to the air. Some are strongly attracted to human corpses. Several species are parasitoids of ants or wasps and show some promise in control of invasive fire ants. May be seen sucking hemolymph (bug blood) from caterpillars or wasps.

A scuttle fly (family *Phoridae*)
1.5 mm.

Puliciphora sp
1 mm.

Family Polleniidae – Cluster Flies

Slightly larger than a house fly; no stripes on thorax; short golden hair. Parasites of earthworms.

Pollenia sp
10 mm.

Family Psychodidae -- Filter flies

Small inhabitants of moist areas, including bathroom and kitchen drains. Not harmful but can be annoying.

Clogmia albipunctata
3 mm.
A.k.a. "bathroom moth midge" or "bathroom moth fly"

Family Rhagionidae – Snipe Flies

Medium-size predator with long legs and substantial piercing mouth parts.

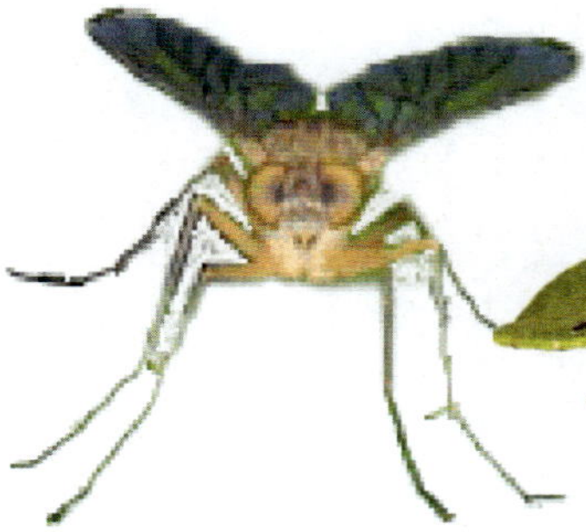

Chrysopilus rotundipennis (F)
4 mm.

Quadrate Snipe Fly
(*Chrysopilus quadratus*) (F)
4 mm.

Family Rhinophoridae – Woodlouse Flies

Small black flies with bristles resembling tachinid flies. Larvae are parasitoids of beetles, spiders, and woodlice.

Melanophora roralis
4 mm.

Families Mycetophilidae and Sciariddae – Fungus Gnats

Mycetophilidae is a large family of small fungus-eating flies, most indistinguishable without microscopic examination. Found in moist areas outdoors, these are not the fungus gnats you find in your greenhouse.

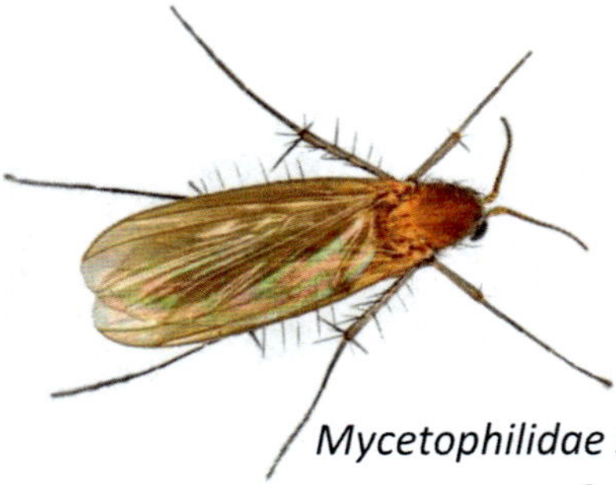

Mycetophilidae sp. – Fungus Gnat
3 mm.

A most vexing pest for indoor gardeners, *Sciarid sp.* fungus gnats seem to spontaneously generate from any pot of dirt. In fact, the larvae were probably already present in the soil when you bought the plant, feeding on roots. The adult flies may live only a week to 10 days, just long enough to mate and lay a couple hundred more eggs in the soil.

Dark-winged Fungus Gnat
(*Sciaridae sp*)
2 mm.

Family Sciomyzidae – Marsh Flies

Also called Snail-killing Flies, as the larva feed on snails.

Limnia sp
6 mm.
Possibly *Limnia Shannonii*

Family Stratiomyinae -- Soldier Flies

2700 species of soldier flies cover diverse terrestrial and aquatic habitats. Typically they are small, often have green or yellow coloration, and rest with their wings stacked. Depending on the species, the adult fly may be a pollen or nectar eater or may not eat at all. Larvae may be predatory or may eat fungi or roots. They look a bit like wasps but will not sting.

Psellidotus hieroglyphicus
7 mm.

Family Syrphidae – Hover Fly

Also called "flower flies," syrphidae adults feed on nectar and pollen, often hovering in flower beds. The larvae may feed on detritus or fungi in the garden; some are important predators of aphids and related garden pests. Many syrphid flies mimic bees or wasps but lack stingers.

Drone Fly
(*Eristalis tenax*)
This mimic of the European honeybee lays its eggs (larvae are "rat-tailed maggots") in animal carcasses.

Transverse Flower Fly
(*Eristalis transversa*) (M)
6 mm.
Larvae consume aphids.

Eupeodes sp.
6 mm.

Sphaerophoria contigua
8 mm.
Larvae feed on aphids, mites, and thrips

Thin-lined Calligrapher
(Toxomerus boscii)
4 mm.

Eastern Calligrapher
(Toxomerus geminatus)
8 mm.

Toxomerus marginatus
5 mm.
Larvae consume aphids, thrips, caterpillars

Family Tabanidae – Horse Flies

Large flies with prominent (often vibrantly-colored) eyes and stout bodies. In females, the eyes are well separated, and the fly is equipped with substantial biting mouth parts. The male has eyes that come together, and weak mouth parts. While the male feeds on nectar and pollen, the female augments her diet with mammal blood needed to produce eggs.

Deer Fly
Chrysops callidus

Black Horse Fly
(*Tabanus atratus*)
20 mm.

Black Horse Fly larva
(*Tabanus atratus*)

Striped Horsefly
(*Tabanas lineola*) (M)
12 mm.

Family Tachimidae – Tachnid Flies

The kids are killers! Tachnid flies lay their eggs on or in insects or arachnids. If the host is lucky, the tachinid species will be a parasite and simply take nourishment from the host. In most cases, however, tachinid larvae are parasitoids; that is, they will devour the host, generally from the inside. Tachnids are generally recognized but their prominent bristles. While some tachinid flies are generalists and will parasitize many different species, others are more selective. For instance, *Istocheta aldrichi*, only attacks the Japanese beetle and can therefore be used as a biological control.

Dexiini sp.

Winthemiini (tribe) sp

Family Tephritidae – Fruit Flies

Fruit flies again? Yes, there are two families called fruit flies. The Drosophilidae are the better known fruit flies; *Tephritidae* are sometimes called peacock flies to avoid confusion and to honor their elaborately patterned wings.

Sunflower Seed Maggot
(*Neotephritis finalis*)
6 mm.

Eutreta rotundipennis
4 mm.

Walnut Husk Maggot
(*Rhagoletis suavis*)

Family Tipulidae – Crane Flies

Sometimes called "mosquito hawk," the crane fly is a nectar eater; its larvae generally feed on algae and aquatic detritus. Harmless to humans and mosquitos alike.

–

Limonia domestica – 7 mm.

Limonia rara (M)

Tipula sp

Tipula Metacomet
40 mm. (body)
Named for the Wampanoag chief Metacomet

Family Ulidiidae – Picture-wing Flies

A large family of detritivores noted for their patterned wings. These small flies are often mistaken for fruit flies, but unlike fruit flies they do not feed on live fruit.

Delphinia picta

7 mm.

A picture-wing fly of no agricultural or medical significance. It is small, attractive, and common in our area. They are relatively placid, sometimes perching in the sun to be admired.

Order Ephemeroptera – Mayflies

Mayflies live most of their life—sometimes years-- as aquatic "naiads." They will molt many times before emerging as a subimago that resembles an adult but not fully developed. The subimago phase may last only a few minutes or many hours before the mayfly molts a final time and the adult emerges. Some species live in their flying form for only a few hours or days. Mayflies emerge at the same time (not necessarily in May) like the cicadas, to overwhelm predators.

Family Baetidae – Small Minnow Mayflies

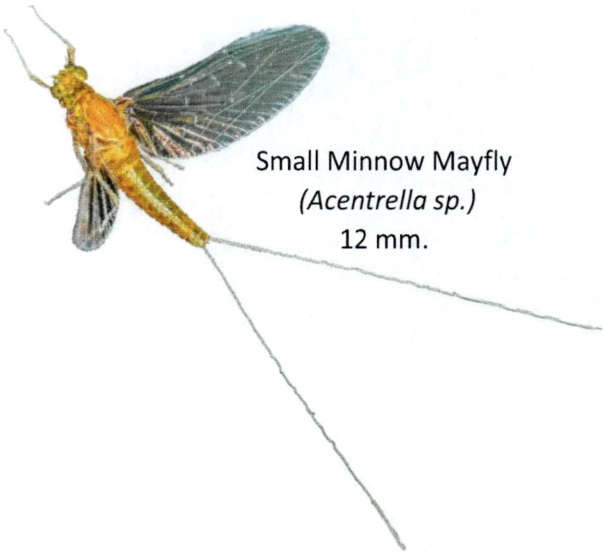

Small Minnow Mayfly
(Acentrella sp.)
12 mm.

Family Ephemerida – Burrower Mayflies

Hexagenia sp.
a Common burrower mayfly
40 mm.

Mayfly exuvia
When the subimago molts, the exoskeleton (exuvia) is left behind.

Spiny Crawler Mayfly
(Ephemerella sp.)
10 mm.

Order Hemiptera – True Bugs, Hoppers, and Aphids

Wait a minute! If these are the true bugs, what are all those spiders and flies? Untrue bugs? The term "bug" has been hijacked and applied to just about any creepy crawly thing without fur or feathers. "Bug" is a short, cute word that is easy to spell, so it's used for all sorts of things, including all the insects. Technically, however, "bugs" are all part of the 100,000+ species that make up the order "Hemiptera. [10] " Each hemipteran has a piercing mouth part (beak) with two channels: one that delivers digestive juices and one that sucks up the resultant nutritional soup from its plant or animal victim. If "piercing mouth part" sounds painful, it can be. Although most bugs are small, vegetarian, and incapable of penetrating all but the most tender human skin, some can deliver a painful "bite." The common and scary-looking wheel bug (*Arilus cristatus)*, for instance, is not reluctant to stab you if you annoy it. When in doubt, leave wheel bugs and related assassin bugs alone.

Family Acanaloniidae – Planthoppers

Acanalonia sp
3 mm.

Acanalonia sp
4 mm.

Cone-headed Planthopper
(*Acanalonia conica)*
6 mm.

---

[10] True bugs were until recently considered their own Order: Heteroptera. They are now generally considered a Suborder of Hemiptera (along with aphids and true hoppers), and all Hemipterans are generally referred to as True Bugs.

Family Achilidae – Planthopper

Catonia sp. – probably *C. Carolina*
5 mm.

Family Alydidae – Broad-headed Bugs
Generally seed-eaters, more often found in arid areas.

Alydus pilosulus
10 mm.

Family Aphididae – Aphids

Oleander Aphid
(*Aphis neriii)*
3 mm.

Purple-spotted Lily Aphid
(*Macrosiphum lilii)*
3 mm.

Woolly Alder Aphid
(Prociphilus tessellatus)
3 mm.

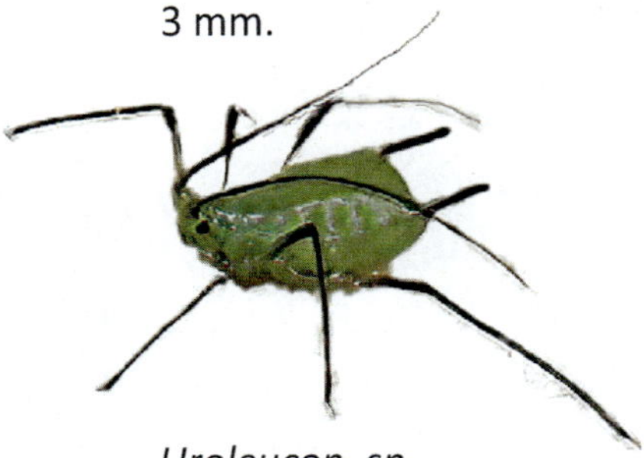

Uroleucon sp
3 mm.

Spittlebugs – Families Aphrophoridae, Cercopidae, and Clastopteridae
Spittlebug nymphs are notable for forming protective balls of foam resembling spit in which they feed and develop before emerging as winged adults (froghoppers).

Family Aphrophoridae

Meadow Spittlebug
(*Philaenus spumarius*)
5-7 mm.

Family Cercopidae

Two-lined Spittlebug
(*Prosapia bicincta*)
9 mm.

Family Clastopteridae

Juniper Spittlebug
(*Clastoptera juniperina*)
4 mm.

Family Aradidae – Flat Bugs

Mezira sayi
12 mm.

Family Belostomatidae – Giant Water Bugs

Eastern Toe-Biter
(*Benacus griseus*)
A Giant Water Bug
65 mm.

Family Cicadellidae – Leafhoppers

Four-spotted Clover Leafhopper
(*Agallia quadripunctata*)
4 mm.

Anoscopus serratulae
4 mm.

Draeculacephala constricta
5 mm.
(a sharpshooter)

Sharp-headed sharpshooter
(*Draeculacephala crassicornis*)

Potato Leafhopper
(*Empoasca fabae*)
4 mm.

Erythroneura comes
4 mm.

Erythroneura calycula
3 mm.

Eratoneura ardens
3 mm.

Erythroneura integra
3 mm.

Grapevine Leafhopper
(*Erythroneura vitis*)
3 mm.

Red-banded Leafhopper
(*Graphocephala coccinea*)
7 mm.

Graphocephala versuta
6 mm.

Gyponini sp.
7 mm.
Probably Eight-lined Leafhopper
(*Gyponana octolineata*)

Gyponini sp. – 1 mm.

Jikradia olitoria nymph
2 mm.
aka Coelidia olitoria

Coppery Leafhopper
(Jikradia olitoria)
6-8 mm.

Aster Leafhopper
(*Macrosteles quadrilineatus*)
4 mm.
aka six-spotted leafhopper

Broad-headed Sharpshooter
(*Oncometopia orbona*)
12 mm.

Japanese Leafhopper
(*Orientus ishidae*)
6 mm.

Paraphlepsius collitus
4 mm.

Ponana pectoralis
6 mm

Scaphoideus sp.
5 mm.

Xestocephalus superbus
4 mm.

Huge by comparison, the cicadas are indeed Cicadellidae

These Brood X cicadas were photographed in Fairfax County. Culpeper won't see *Magicicada* cicadas until Brood II returns in 2030.

Pharaoh Cicada
(*Magicicada septendecim*)
32 mm.
17-year cicada, Brood X

Cassin's Periodic Cicada
Magicicada cassinii
17-year cicada, Brood X
24-26 mm.

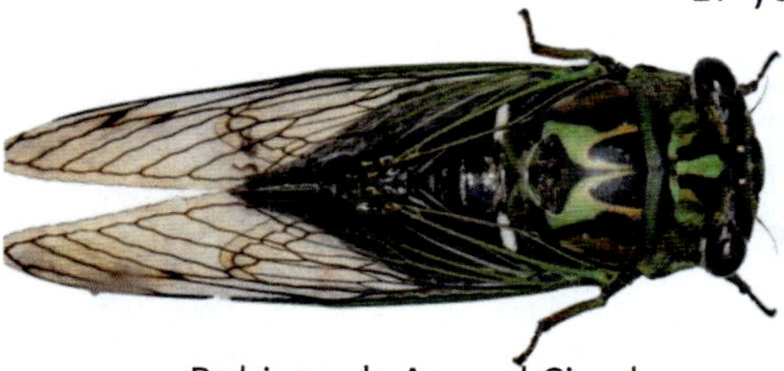

Robinson's Annual Cicada
(*Neotibicen robinsonianus*)
27 mm.

Swamp Cicada
(*Neotibicen tibicen*)
27 mm.

Family Coreidae – Leaf-footed Bug

Acanthocephala terminalis.
4 mm.

Acanthocephala terminalis nymph

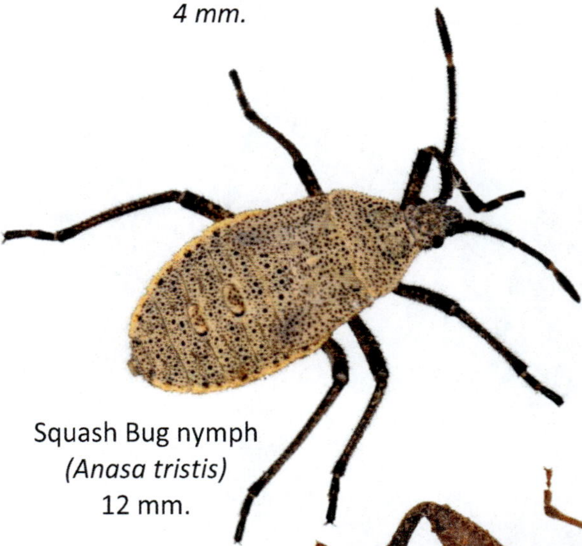

Squash Bug nymph
(Anasa tristis)
12 mm.

Piezogaster sp.
(possibly *P. calcarator)*
17-21 mm.

Family Corixidae – Water Boatmen

Trichocorixa sexcincta
or
Hesperocorixa linnaei
6 mm.

Family Cydnidae – Burrower Bug

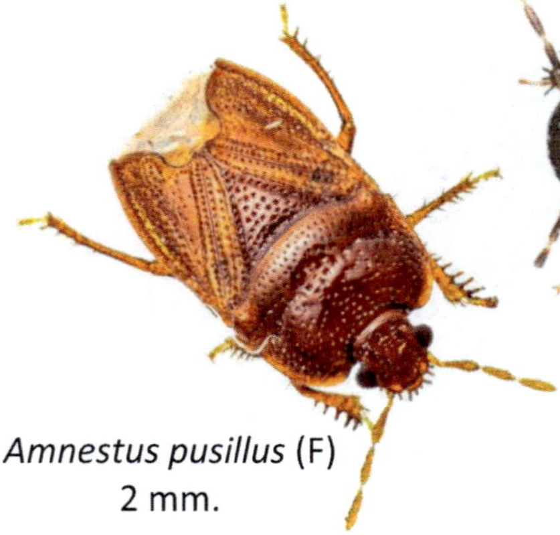

Amnestus pusillus (F)
2 mm.

Two-lined Burrowing Bug
(*Pangaeus bilineatus*)
aka Peanut Burrower Bug
5-7 mm.

White-margined Burrower Bug
(*Sehirus cinctus*)
4-6 mm.

Family Delphacidae – Delphacid Planthopper

*Megamelus davis*l
5 mm.
Congregating on waterlily pads

Family Derbidae – Derbrid Planthopper

Anotia bonnetii
8 mm.

Family Flatidae – Flatid Planthoppers

Northern Flatid Planthopper
(*Flatormenis proxima)*
9 mm.

Citrus Flatid Planthopper
(*Metcalfa pruinosa)*
6-8 mm.
Adult and waxy nymph

Family Issidae -- Issid Planthopper

Aplos simplex nymph
3 mm.

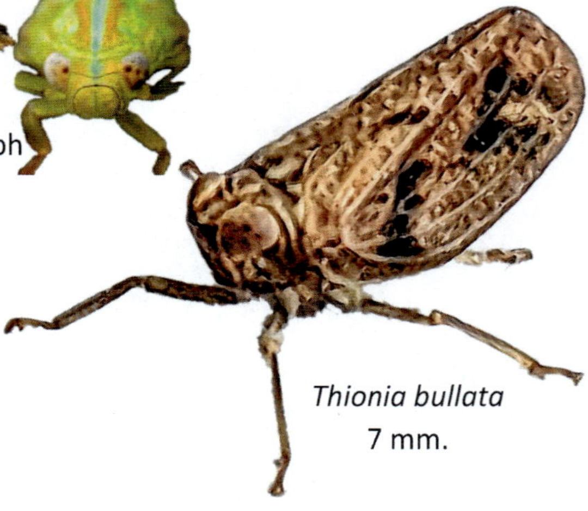

Thionia bullata
7 mm.

Family Lydaeidae – Seed bugs

Sycamore Seed Bug
(*Belonochilus numenius*)
9 mm.

Small Milkweed Bug
(*Lygaeus kalmi*)
10 mm.

Whitecrossed Seed Bug
(*Neacoryphus bicrucis)*
7-10 mm.
(a.k.a. Ragwort Seed Bug)

Large Milkweed Bug
(*Oncopeltus fasciatus*)
10-18 mm.

Family Membracidae – Treehoppers

Membracids are small insects that suck sap from plants and excrete the excess sugar as honeydew. In most species the honeydew attracts ants who consume the honeydew and provide protection for the treehoppers. These bugs are frequently gregarious, with dozens found near each other. This concentration of insects attracts more ants. Membracids may lay their eggs on a stem and guard them as well as the nymphs. The ants provide protection for all and "farm" the treehoppers: Ants will stroke the treehoppers to induce honeydew excretion.

Two-marked Treehopper
(Complex *Enchenopa binotata)*
4 mm.

Enchenopa binotata is a "complex" because it consists of numerous identical species that are best distinguished by their host plant. This individual was on black walnut.

Enchenopa sp.
2 mm

An undescribed/unnamed treehopper that feeds on walnut or butternut . This insect is not uncommon but has not yet been formally described and assigned a species name. I have also seen this in a green and white version.

Keeled Treehopper
(*Entylia carinata*)
Adult (5 mm.), eggs, & nymph.

← Ants attending treehopper and nymphs.

↙

Ceresini tribe nymph
4 mm.

Honeylocust Treehopper
(*Micrutalis calva)*
3 mm.

Oak Treehopper
(*Platycotis vittata*) .
4 mm.

A Buffalo Treehopper
(*Stictocephala sp.)*
3 mm.

Telamona extrema
8 mm.

Family Miridae – Plant Bug

The largest family of "true bugs" with more than 10,000 described species. These small bugs which include numerous agricultural pests are generally sap suckers, although some are predatory.

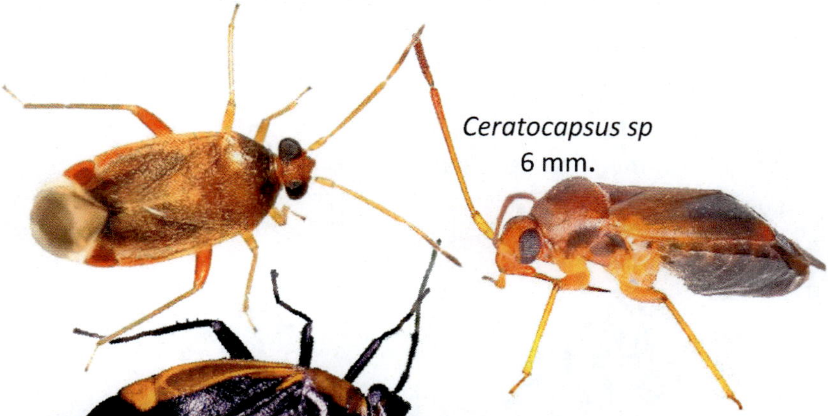

Ceratocapsus sp
6 mm.

Metriorrhynchomiris dislocatus
10 mm.

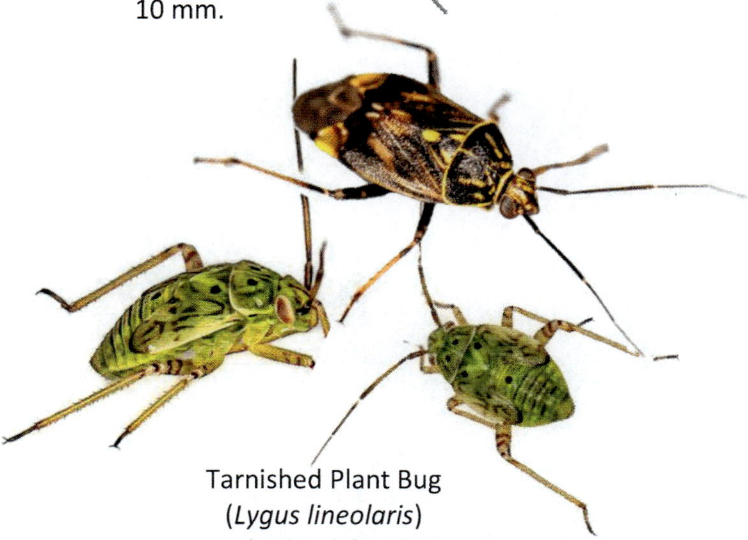

Tarnished Plant Bug
(*Lygus lineolaris*)
nymphs & adult -- 5 mm

L. lineolaris is a full-spectrum pest, feeding on about half of all agricultural crops, and a major pest of loblolly pine seedlings.

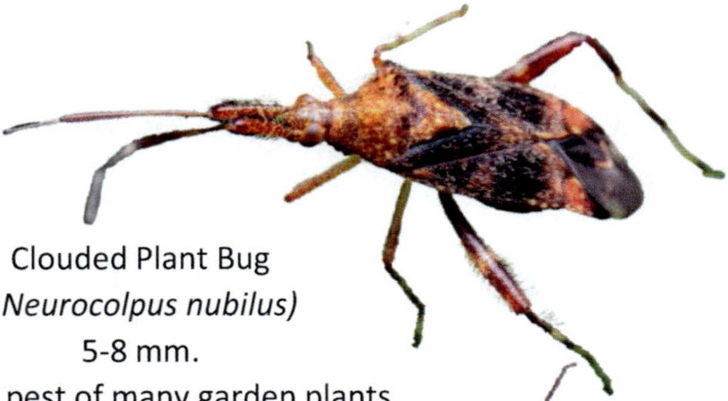

Clouded Plant Bug
(*Neurocolpus nubilus)*
5-8 mm.
Another pest of many garden plants.

Phytocoris sp
5 mm.

Pycnoderes sp
3 mm.

Ornate Plant Bug
(Reuteroscopus ornatus)
4 mm.

Sericophanes heidemanni
4 mm.

Two-spotted Grass Bug
(*Stenotus bidotatus*)
6 mm.

Broken-backed Bug
(*Taylorilygus apicalis*)
3 mm.

Rice Leaf Bug
(*Trigonotylus caelestialium*)
5 mm.

Family Rhopalidae – Scentless Plant Bug

So named because of their very small scent gland opening...not something you'll notice

Harmostes fraterculus
5 mm.

Hibiscus Scentless Plant Bug
(Niesthrea louisianica)
6-8 mm.

N. louisianica eggs

Family Pachygronthidae

Primarily seed and sap-feeding bugs, with a few predatory species.

Oedancala dorsalis
6 mm.

Phlegyas abbreviates
5 mm.

Family Pentatomidae – Stink Bugs

More than 200 species in North America. The entire family gets a bad reputation because of the brown marmorated stink bug's tendency to invade houses in large numbers.

Juniper Stink Bug
(*Banasa euchlora*)
9-11 mm.

Green Stink Bug
(*Chinavia hilaris*)
13-19 mm.

Twice-stabbed Stink Bug
(*Cosmopepla lintneriana*)
5-7 mm.

Dusky Stink Bug
(*Euschistus tristigmus*)
8-11 mm.

The Florida predatory stink bug is a species of carnivorous shield bug. It is native to the hottest parts of the southeastern United States and is considered beneficial because it feeds on many species of pest insects. As average temperatures rise, *E. floridanus* has been expanding its range north, arriving in Virginia just a few years ago.

Florida Predatory Stink Bug
(*Euthyrhynchus floridanus*)
12-17 mm.

--

Brown Marmorated Stink Bug
(*Halyomorpha halys*)
12-17 mm.

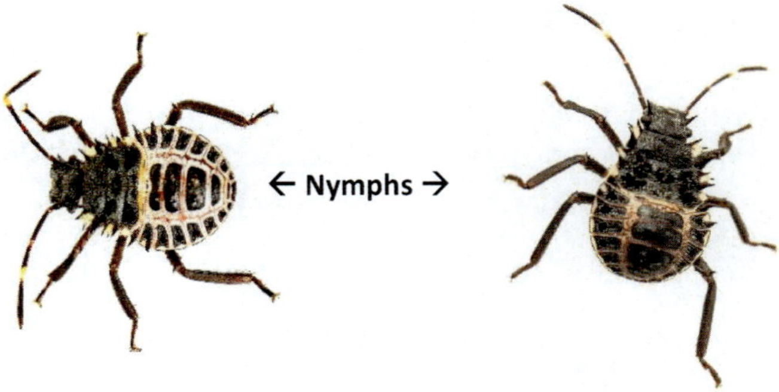

Yes, this is the one you love to hate! These little stinkers love to spend cold weather in warm houses like yours. In Fall and Spring, you'll find them crawling on your windows and buzzing around your living room lights. When caught or smashed they exude an unpleasant and lingering odor. Outdoors it does significant damage to fruit crops. Now common in our region it is a recent invasive, first noted in Pennsylvania around 2000.

Mormidea lugens
5-7 mm.

Rice Stink Bug
(*Oebalus pugnax*)
8-12 mm.

Family Plataspidae

Although this bug prefers to lunch on kudzu, it will also eat soybeans and related plants. Will it harm me? The bite is reported to cause some patches of skin to die in sensitive individuals, though no permanent harm has been noted.

Kudzu Bug
(*Megacopta cribraria*)
4 mm.

Family Reduviidae – Assassin Bugs

Wheel Bug
(*Arilus cristatus*)
38 mm.

Wheel Bug nymph on egg mass

Caution: The wheel bug is a common beneficial insect in most gardens, but it is a large bug with a formidable beak. Although generally docile, if mishandled or threatened the adult and nymphs will deliver a painful "bite." The wound is said to be more painful than a wasp sting and may take weeks to heal. If you want to handle this bug, wear gloves.

Jagged Ambush Bug
(*Phymata sp.*)
7 mm.
on Swamp Milkweed

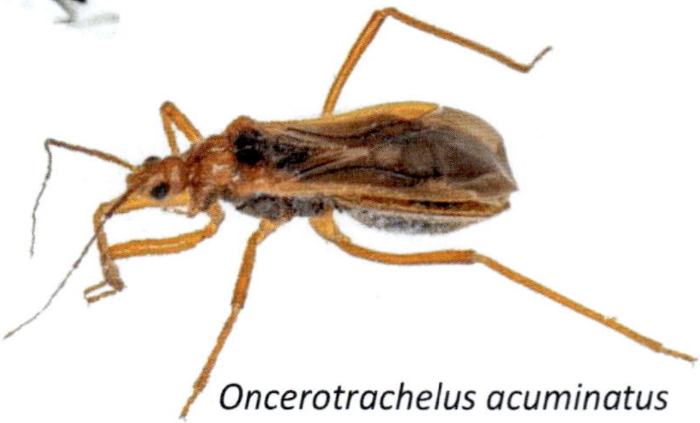

Oncerotrachelus acuminatus
6 mm.

Pnirontis sp
10-12 mm.

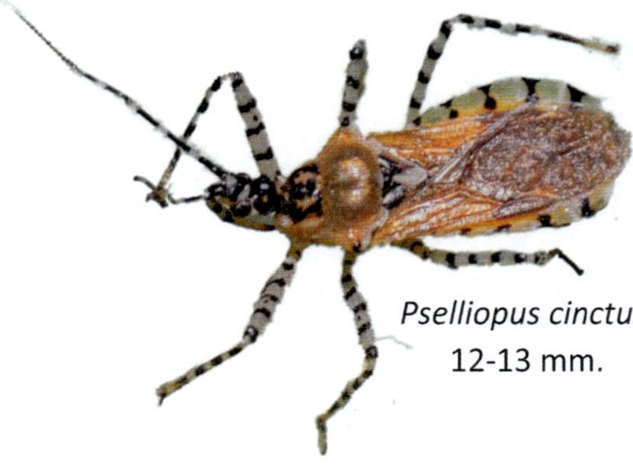

Pselliopus cinctus
12-13 mm.

Spiny Assassin Bug
(Sinea spinipes)
12-14 mm.

Sinea sp. nymph
3 mm.

Sirthenea carinata
20-25 mm.

Stenopoda spinulosa
22-30 mm.

Zelus luridus
12-18 mm.

Family Rhyparochromidae – Dirt-colored seed bugs

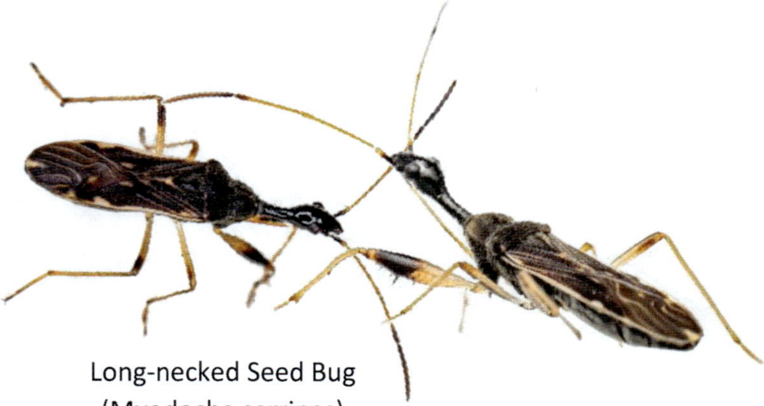

Long-necked Seed Bug
(Myodocha serripes)

Ptochiomera nodosa

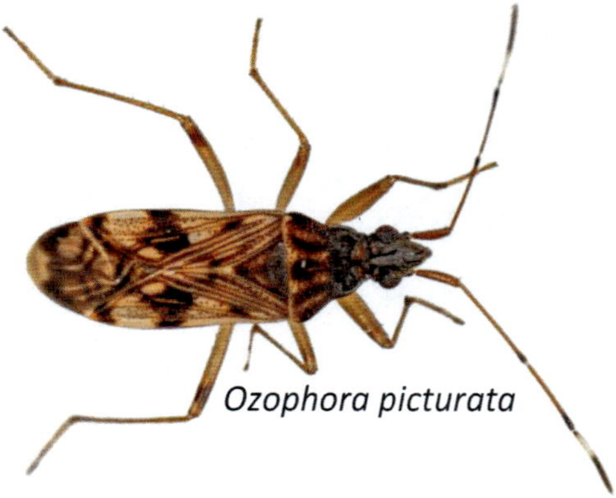

Ozophora picturata

Family Thyreocoridae – Ebony Bug

Corimelaena sp.

Family Tingidae – Lace Bugs

Walnut Lace Bug
(*Corythucha juglandis*)
4 mm.

Lace Bug nymph
1 mm.

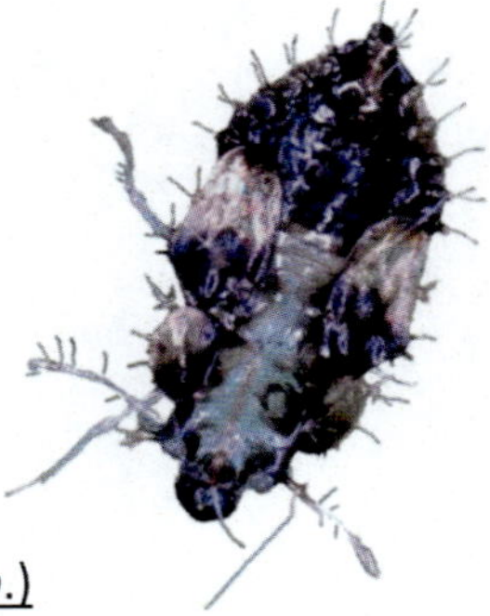

Lace Bug
(*Tingidae sp.*)
3 mm.

Order Hymenoptera – Bees/Wasps/Ants/Saw Flies

Family Formicidae – Ants

Reddish Carpenter Ant
(*Camponotus castaneus*)
7 mm.
Tending Keeled Leafhoppers (adult and nymphs). Nests in rotting logs, soil under rocks, etc., or even in exposed soil.

Black Carpenter Ant
(*Camponotus pennsylvanicus*)
6-7 mm smallest minor worker; 12-14 mm large major worker; 15-18 mm queen. The nest is in dead, usually already rotten wood. Mating flights are usually on the first warm humid afternoons in spring.

Crematogaster sp.
an acrobat ant

Crematogaster ashmeadi
an acrobat ant
3 mm.

Acrobat ants are arboreal, nesting and foraging in trees.

Formica (fusca group) – field ants

Sun-loving, often found in soil in open grassland, meadows, and sandy open woodlands, even in lawns. Tends treehoppers, etc.

Formica subsericea
a field ant

Pale Field Ant
(Formica pallidefulva)

American Cornfield Ant
(Lasius americanus)
2-4 mm.
Found in shaded woodland, where it nests in rotting logs and stumps and under stones.

False Honey Ant
(Prenolepis impairs)
aka Winter Ant being eaten by an Emerald Jumping Spider

Strumigenys rostrata
a miniature trap-jaw ant
2.5 mm.

Odorous House Ant
(*Tapinoma sessile*)
3 mm.

If you have small ants in your kitchen, there's a good chance this is the one. Also called "coconut ant," it has a distinct odor when crushed. Offering them a bottle cap with sugar and boric acid will likely eliminate them.

Family Argidae – Sawflies

Hibiscus sawfly
(*Atomacera decepta*) larva
13 mm.

Hibiscus sawfly
(*Atomacera decepta*)
4-8 mm.

Butternut Wollyworm
(Eriocampa juglandis)
A pest of walnuts (*Juglans* is the plant species of walnut)

Allantinae sp. (larva)

Family Andrenidae – Mining Bees

Andrina sp.

Family Apidae – Carpenter Bees

Eastern Carpenter Bee
(Xylocopa virginica) male – 24 mm.

Will it harm me? Not normally, although it will harass you. The male (non-stinging) guards its territory and will buzz you in an annoying manner. It can, however, damage your house by boring into wood. The damage may be compounded by woodpeckers hammering their way into the wood to eat the bee larvae.

Family Halictidae – Sweat Bees

Slender, tiny to medium bees, often dusky black, dull green or blue, with bands of hair on their abdomen.

Bicolored Striped Sweat Bee
(Agapostemon virescens)
10 mm.

Metallic Epaulleted Sweat Bee
(Augochloropsis metallica)
9 mm.

Ligated Furrow Bee
(*Halictus ligatus*)
9 mm.

Lasioglossum sp
4 mm.

Identifying *Lasioglossum* to species is difficult because they look similar. Often the differences between species is the density of punctuations on the thorax and the surface texture between the punctuations.

Family Trigonalidae

Trigonalids are hyperparasitoids: That is, they parasitize parasitoids.

Taeniogonalos gundlachii
8 mm.

Per Ross Hill, this is "...a parasitoid of ichneumonid wasps and tachinid flies. Females lay their eggs along the upper edge of a leaf that are then eaten by a feeding caterpillar. After ingestion of the eggs, if the caterpillar is parasitized by a primary host (a wasp or fly), upon hatching, the trigonalid larvae seek out the host larvae and parasitize them. They then emerge from the primary host puparium."[11]

That seems to be a pretty "iffy" proposition but must work since this bee is common across the U.S.

---

[11] Quoted from BugGuide.net/node/view/132680. They explain it much better than I could.

Family Braconidae – Braconid wasp

The braconid wasps, like the closely-related ichneumon wasps, are parasitic. They lay their eggs on or in other insects or arthropods, where the larvae develop and usually kill the host.

Smooth-headed Mummy Wasp
(*Aleiodes politiceps*)
10 mm.

Earinus limitaris
7 mm.

Cotesia congregate eggs on Tobacco Hornworm

Family Chrysididae – Cuckoo Wasp

Cuckoo wasps are kleptoparasites: They lay their eggs on the nests of unrelated wasp species, where their larvae eat the hosts larvae and then eat the food provided by the host. Cuckoo wasps are generally selective when it comes to hosts, often parasitizing only one species.

Cuckoo Wasp
(Chrysis angolensis)
12 mm.
Lays its eggs on mud dauber nests.

Family Crabronidae – Square-headed wasps

Cerceris halone
12 mm.
Nests in burrows, providing weevils for its larvae.

Tachytes guatemalensis
19 mm.
Provisions its burrows with grasshoppers and katydids.

Hump-backed Beewolf
(Philanthus gibbosus)
9 mm.
Frequently preys on bees.

Family Ichneumonidae – Ichneumon wasp

Parasitic wasps that resemble stinging wasps but with longer antennae, many of which have middle segments that are yellowish or whitish; females may have extremely long ovipositors. Very few Ichneumonids sting, and those that sting do not inject poison. The ichneumonids are endoparisites: They inject single eggs into a host (egg, larva, or pupa). The larval wasp eats its way out, killing the host. Some ichneumonids kill or paralyze the host when the egg is deposited. Other allow the host to develop into a pupa before it is eaten.

Cratichneumon sp
11 mm.

Dolichomitus irritator (F)
22 mm.

Enicospilus purgatus
25mm.

Ophion sp.
18 mm.

Limonethe maurator
12 mm.

Family Multillidae – Velvet Ants

Parasitoids of ground-nesting wasps and bees. Female velvet ants will search out wasp nests and lay an egg in a host pupa.

A velvet ant (*Dasymutilla scaevola*) female
7 mm.
a.k.a. "Cow killer"

Females wingless, very hairy, and may look like large ants but have no node (bump) on "waist" between abdomen and thorax (ants have one or two).

Males winged, less hairy, look more like typical wasps, larger than females

Caution: Although called an ant, this is a wasp. The male has a non-stinging "pseudo stinger," but the female has a real stinger and will deliver an exceptionally painful sting if handled.

Family Pompilidae – Spider Wasps

Larvae feed on spiders. In most species the females sting and paralyze their prey and then transport it to a specially constructed nest before laying an egg on the spider. When the egg hatches the larva will feed on the paralyzed spider.

Auplopus mellipes
11 mm.

Rusty Spider Wasp
(Tachypompilus ferrugineus ferrugineus)
26 mm.

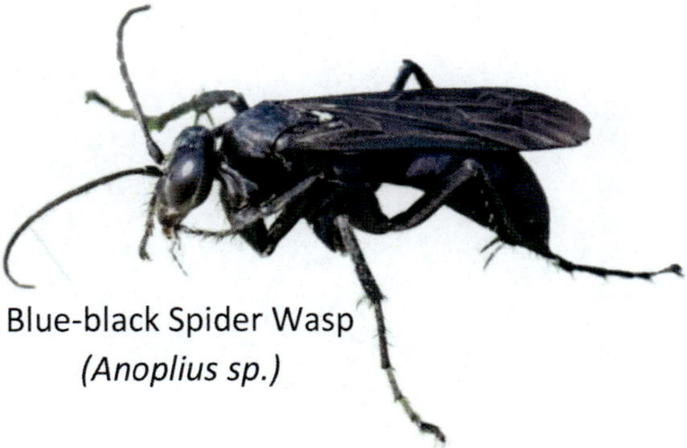

Blue-black Spider Wasp
(Anoplius sp.)

Family Scelionidae – Parasitic Wasps

Very small wasps that parasitize insect eggs.

Gryon (pennsylvanicum?) wasp parasitizing Squash Bug eggs
The wasp is about 2 mm.

Trissolcus sp. wasp parasitizing *Podisus* eggs
The wasp is about 2 mm., slightly larger than the stinkbug eggs.

Family Scoliidae – Scolid Wasps

Blue-winged Wasp
(*Scolia dubia*)
24 mm.

Also known as Digger Wasp, males and females have a courtship dance, flying close to the ground in a figure-8 or S pattern. Females burrow into ground in search of grubs, especially those of green June beetles and Japanese beetles She stings it and often burrows farther down, then constructs a cell and lays an egg on the host. Larva pupates and overwinters in a cocoon within the body of the host. One generation per year in North, more in South.

Family Sphecidae – Thread-wasted Wasps

Ammophila sp

20-25 mm.

Commonly found on wildflowers with large clusters of blossoms, such as Queen Anne's Lace. Mating pairs on flowers are common.

Eremnophila aureonotata

22 mm.

Adults take nectar; larvae feed on larval butterflies and moths. Female digs a burrow and provisions it with a single large caterpillar.

Isodpntia sp.
a grass-carrying wasp

Katydid Wasp
(*Sphex nudus)*
23 mm.

Polistes parametricus
a paper wasp
20 mm.

Family Vespidae – Hornets, Potter Wasps

Females lay egg in mud nest (built on twig, etc.), then provision with small caterpillars, especially cankerworms. Also reported to provision with sawflies.

Eumenes fraternus
a potter wasp
15-20 mm.

European Hornet
(*Vespa crabro*)
24 mm.
Large but not aggressive.

Southern Yellowjacket
(*Vespula squamosa*)
15 mm.

Common and aggressive, the Southern Yellowjacket fills an important role in the environment but is best kept out of your yard. Yellowjackets nest underground in large colonies, but "underground" may mean in your large outdoor flowerpots or under your hardscaping. If disturbed, this wasp will defend its territory against all insects, pets, and people. Funnel traps baited with meat may effectively deplete the colony, and the Internet is loaded with ideas for removing this pest from your yard. Some of those ideas are dangerous. When in doubt, consult your extension agent.

Order Lepidoptera – Skippers, Butterflies, & Moths

Moths, butterflies, and skippers all have membranous wings covered with pigmented scales, for which they are given their taxonomic name, "Lepidoptera," meaning "scaly wings." In general, butterflies are more vibrantly colored and operate during the day while moths rule the night and are rather dull, but there are many exceptions. Some moths are quite colorful, and a walk through an open field most days will stir up a multitude of small Crambid moths. More reliably, butterflies have clubbed antennae (knobs on the end) while moths have straight or filiform antennae without clubs. (Male moths often have quite elaborate antennae.) Skippers are an intermediate design looking more like a moth but with antennae that are clubbed and include a hook on the end. Butterflies emerge from solid chrysalides, while moths emerge from silken cocoons. There are exceptions to all these generalizations. Perhaps the most reliable distinction is that the forewing and hindwing of moths are connected by a frenulum and operate together; butterfly wings are not connected. Skippers usually lack the frenulum as well.

Butterflies

Family Hesperiidae – Skippers

Larvae feed on various grasses, including Bermuda Grass, crabgrass, St. Augustine Grass, and goosegrass. Adults take nectar from flowers of many herbaceous and shrubby plants.

Sachem Skipper
(*Atalopedes campestris*) ♂
35-41 mm.

Silver-spotted Skipper
(*Epargyreus clarus*)
45-67 mm.

Horace's Duskywing caterpillar
(*Erynnis horatius)*

Female Zabulon Skipper Male
(*Poanes zabulon* or *Lon zabulon*)
35-41 mm.

Common Checkered Skipper
(*Pyrgus communis)*
19-32 mm.

Peck's Skipper
(*Polites peckius***)**
19-27 mm.

Family Lycaenidae – Blues, Coppers, Hairsreaks

Caterpillar eats flowers and seeds of a wide variety of native and imported herbs and occasionally shrubs, including White Clover, Red Clover, and Cow Vetch. This species has thrived where its habitat has been encroached upon by human activities. It is common along freshly mowed roadsides, flying to puddles.

Summer Azure
(*Celastrina neglecta***)**
23-29 mm.

Eastern Tailed-Blue
(*Cupido comyntas*)
23-28 mm.

Males perch all afternoon on small trees and shrubs to seek receptive females. Eggs are laid singly on flowers of host plant. Young caterpillars feed on flowers and fruits; older ones may eat leaves. Caterpillars are sometimes attended by ants, which receive a sugary solution from the dorsal nectary organ

American Copper
(*Lycaena phlaeas)*
22-35 mm.

Lycaenidae sp. (poss. Hairstreak)

Gray Hairstreak
(*Strymon melinus*)
22-35 mm.

Family Nymphalidae – Brush-footed butterflies

Hackberry Emperor
(*Asterocampa celtis*)
48-64 mm.

Tawny Emperor
(*Asterocampa clyton*)
50-72 mm.

Silvery Checkerspot
(*Chlosyne nycteis*)
40-52 mm.

Successive instars of the Monarch caterpillar differ only in size

When the caterpillar is ready to pupate, it hangs upside down in a "J" shape for many hours before suddenly emerging as a chrysalis. It remains a chrysalis for about twelve days.

24-hours before the butterfly is ready to emerge (eclose), the green chrysalis shell becomes transparent.

Monarch
(*Danaus plexippus*) male
86-125 mm.

Variegated Fritillary
(*Euptoieta claudia*)
44-80 mm.

Common Buckeye
(*Junonia coenia*)
40-68 mm.

Northern Pearlyeye
(*Lethe anthedon*)
53-67 mm.

American Snout
(*Libytheana carinenta*)
41-51 mm.

Viceroy caterpillar
4th instar

Viceroy
(*Limenitis archippus*)
67-85 mm.

Red-spotted Purple
(*Limenitis arthemis*)
75-101 mm.

Little Wood Satyr
(*Megisto cymela*)
39-47 mm.

Pearl Crescent
(*Phyciodes tharos*)
35 mm.
Larval foodplant: Asters. Adults take nectar

Almost identical to the Northern Cresent (*Phyciodes cocyta*); most of the field identification guidelines are full of exceptions. In general, Northern Crescents are larger than Pearl Crescents, but unless you see them together, that is of little help. Until late in the 20th Century these were considered a single species.

Eastern Comma
(Polygonia comma)
45-64 mm.

Question Mark
(*Polygonia interrogationis*)
57-75 mm.

Painted Lady
(Vanessa cardui)
51-73 mm.

Red Admiral
(Vanessa atalanta)
45-76 mm.

Family Papilionidae – Swallowtails

Eastern Tiger Swallowtail
(Papilio glaucus)
62-115 mm.
Female has both yellow and black forms.

Eastern Black Swallowtail
(Papilio polyxenes)
80-110 mm. (adult)

Eastern Black Swallowtail Caterpillar
The orange projections on the head are osmeteria, defensive organs that extend and exude an unpleasant odor when the cat feels threatened.

Family Pieridae – Whites

Falcate Orangetip
(*Anthocharis midea*)
35-45 mm.

Orange Sulphur
(*Colias eurytheme*)
35-70 mm.

Female

Male

Cabbage White
(*Pieris rapae*)
45-58 mm.

Our most abundant butterfly, flying from early Spring until late Fall. This involves numerous generations, as the adult lives for only a few weeks.

Moths

While you may be familiar with the numerous butterflies that frequent your garden, they are the junior partners of the order Lepidoptera. For every butterfly species in the world there are nine moth species: 17,500 butterflies to 160,000 moths. There are many generalizations about the differences, but numerous exceptions to those generalizations.

- Butterflies fly during the day; moths fly at night. Mostly true, but there are many moths that fly during the day as well.
- Butterflies are colorful; moths are drab. Often true, but there are some quite colorful moths.
- Butterflies emerge from hardened chrysalis, moths from a silken cocoon. Again, usually true but there are some exceptions.
- Butterflies have long straight antennae; moths have elaborate, frilly antennae. Some moths (particularly female) have very plain (filiform) antennae, but lack the terminal club characteristic of the butterflies.
- Butterflies perch with their wings folded up; moths leave their wings folded out. Generally true, but butterflies will open their wings when drying or thermal regulating.

Two discriminators that are more reliable are the wing and eye structures. The wings of a moth are connected by a frenulum, so they operate together. Butterfly wings are separate. The eyes of a moth are optimized for low light, but don’t pick up much detail. Butterflies have eyes that combine many images to provide a detailed representation.

Family Apatelodidae – Apatelodid Moths

Spotted Apatelodes Moth
(Apatelodes torrefacta)
35-42 mm.

Family Attevidae – Tropical Ermine Moths

Ailanthus Webworm
(*Atteva aurea*)
18-30 mm.
Quite common and easily spotted in our area.

Family Cossidae – Cossid Millers

Carpenterworm Moth
(*Prionoxystus robiniae*)
40-80 mm.
aka Locust Borer

Family Coleophoridae – Casebearing Moths

The larvae of Coleophoridae build silken cases around themselves for protection, discarding the cases and building new ones as they grow.

Streaked Coleophora Moth
(*Coleophora cratipennella)*
9 mm.

Family Crambidae – Crambid Snout Moths

Although "moths fly at night," walk through a field of grass any warm afternoon and numerous crambid moths will fly as you go by. They will stay close to the ground and disappear into the grass a few feet away.

Invisible Crambid Moth
(*Aethiophysa invisalis*)
9 mm.

Yellow-spotted Webworm Moth
(*Anageshna primordialis*)
12-15 mm.

Pegasus Chalcoela Moth
(*Chalcoela pegasalis*)
8 mm.

Zebra Conchylodes Moth
(*Conchylodes ovulalis*)
23-30 mm.

Double-banded Grass Veneer
(*Crambus agitatellus*)
17-22 mm.

Pasture Grass Veneer
(*Crambus saltuellus*)
22 mm.

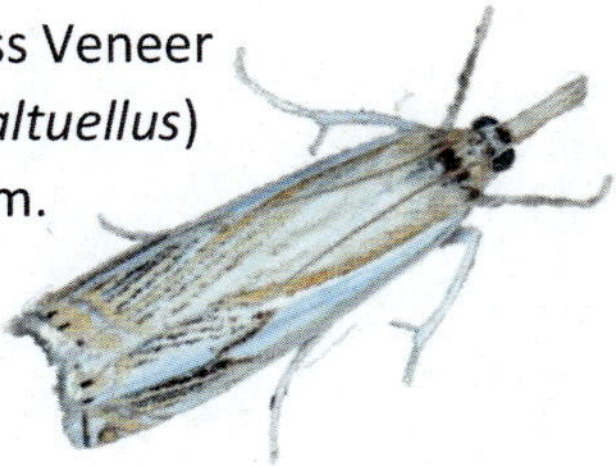

Sawtoothed Crocidophora
(*Crocidophora serratissimalis*)
18-25 mm.

Pale-winged Crocidophora
(*Crocidophora tuberculalis*)
15 mm.

Grape leaffolder
(*Desmia funeralis*)
21-28 mm.

White-headed Grape Leaffolder
(*Desmia maculalis*)
20-26 mm.

Darker Diacme Moth
(*Diacme adipaloides*)
18 mm.

Pickleworm
(*Diaphania nitidalis*)

Harlequin Webworm
(*Diathrausta harlequinalis*)
13 mm.

Julia's Dicymolomia Moth
(*Dicymolomia julianalis*)
8-11 mm TL

Waterlily Leafcutter
(*Elophila obliteralis*)
10-20 mm

Waterlily Borer Moth
(*Elophila gyralis*)
16-30 mm.

Eudonia heterosalis
13 mm.

Striped Eudonia Moth
(Eudonia strigalis)
13-17 mm.

Spotted Peppergrass
(*Eustixia pupula*)
17 mm.

Cross-striped Cabbageworm
(*Evergestis rimosalis*)

Larvae feed on members of the cabbage family, including cabbage, collard greens and Brussels sprouts.

Changeable Grass-Veneer
(*Fissicrambus mutabilis*)
17 mm.

Black-patched Glaphyria
(*Glaphyria fulminalis*)
11 mm.

White-roped Glaphyria Moth
(*Glaphyria sesquistrialis*)
15 mm.

Hahncappsia sp.
24 mm.

Bold-feathered Grass Moth
(*Herpetogramma pertextalis*)
33 mm.

Spotted Beet Webworm
(*Hymenia perspectalis*)
16-22 mm.

Alfalfa Webworm Moth
(*Loxostege cereralis*)
13-25 mm.

Elegant Grass-veneer
(*Microcrambus elegans*)
12-15 mm.

Lucerne Moth
(*Nomophila nearctica*)
24-35 mm.

Obscure Psara
(*Psara obscuralis*)
23 mm.

Herbivorous Pleuroptya Moth
(*Patania silicalis*)
24-26 mm.

Mint-loving Pyrausta
(*Pyrausta acrionalis*)
14-18 mm.

Variable Reddish Pyrausta
(*Pyrausta rubricalis*)
15 mm.

Double-striped Scoparia Moth
(*Scoparia biplagialis*)
6-8 mm. moss-eater

Carrot Seed Moth
(*Sitochroa palealis*)
24-36 mm.

Hawaiian Beet Webworm
(*Spoladea recurvalis*)
22-24 mm.

Celery Leaftier
(*Udea rubigalis*)
10-13 mm. TL

Snowy Urola
(*Urola nivalis*)
15-23 mm.

Family Depressariidae – Flat-bodied moths

Antaeotricha sp.
probably Schlaeger's Fruitworm
*(*Antaeotricha schlaegeri)
21-30 mm.

Dotted Leaftier Moth
(Psilocorsis reflexella)
14-17 mm.

Scythropiodes issikii
a.k.a Odites issikii
14-19 mm.

Family Drepanidae – Hooktip Moths

Arched Hooktip
(Drepana arcuate)
24-40 mm.

Rose Hooktip
(Oreta rosea)
25-34 mm.

Family Erebidae

Now **subfamily Arctiinae** within the family Eribidae, the tiger moths were formerly classified as a separate family (Arctiinae.*)*

Harnessed Tiger Moth
(*Apantesis phalerata*)
30-42 mm.

Banded Tiger Moth
(*Apantesis vittata*)
32-42 mm.

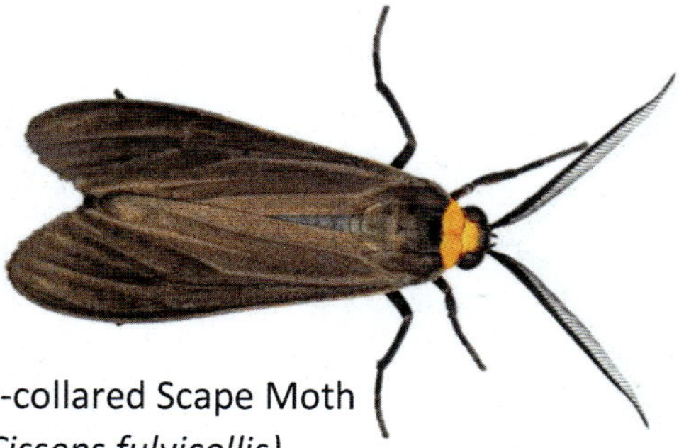

Yellow-collared Scape Moth
(*Cisseps fulvicollis)*
30-37 mm.

Unexpected Cycnia
(*Cycnia collaris*)
33-45 mm.

Delicate Cycnia
(*Cycnia tenera*)
a.k.a. Dogbane Tiger Moth
30-40 mm.

Milkweed Tussock Moth
(*Euchaetes egle*)

Salt Marsh Moth
(*Estigmene acrea***)**

Banded Tussock Moth (*Halysidota tessellaris*)
22-25 mm.
or
Sycamore Tussock Moth (Halysidota harrisii)

Note: These two species can only reliably be separated by dissecting and inspecting the genitalia. All bugs in this book were photographed alive and were unharmed.

Giant Leopard Moth
(*Hypercompe scribonia*)
33-35 mm.

Isabella Tiger Moth
(*Pyrrharctia isabella*)
24-33 mm.

Virginia Tiger Moth & caterpillar
(*Spilosoma virginica*)
17-26 mm.

Agreeable Tiger Moth
(*Spilosoma congrua*)
15-24 mm. TL

False Underwing
(*Allotria elonympha*)
38-50 mm

Hibiscus Leaf Moth caterpillar
(*Anomis private)*

Bent-winged Owlet
(*Bleptina caradrinalis*)
22-32 mm.

Forage Looper Moth
(*Caenurgina erechtea*)
30-42 mm.

Clover Looper
(*Caenurgina crassiuscula*)
dark/light versions
30-40 mm.

Ilia Underwing
(*Catocala ilia*)
65-82 mm.

Ultronia Underwing
(*Catocala ultronia*)
46-60 mm.

Pale Lichen Moth
(*Crambidia pallida*)
19-25 mm.

Morbid Owlet
(*Chytolita morbidalis*)
29-35 mm.

Yellow-lined Owlet
(*Coloibochyla interpuncta*)
20-24mm.

Red-footed Hypena
(*Hypena eductalis*)
25 mm.
a.k.a. Red-footed Bomolocha, Red-footed Snout, or Alder Smoke Moth

Gray-edged Hypena
(*Hypena madefactalis*)
25-32 mm.

Flowing-line Snout
(*Hypena manalis*)
23-28 mm.

Baltimore Snout
(*Hypena baltimoralis*)
26-32 mm.

Green Cloverworm moth
(*Hypena scabra*) - 19 mm
a.k.a. Black snout moth

Black-patched Graylet
(*Hyperstrotia secta*)
17 mm.

White-lined Graylet
(*Hyperstrotia nana*)
18 mm.

Fall Webworms
(*Hyphantria cunea*)

Painted Lichen Moth
(*Hypoprepia fucosa*)
25-35 mm.

Small Necklace Moth
(*Hypsoropha hormos*)
25-34 mm.

American Idia Moth
(*Idia americalis*)
20-30 mm.

Common Idia Moth
(*Idia emula*)
20-30 mm.

Ambiguous Moth
(*Lascoria ambigualis*)
21-25 mm.

Lost Owlet moth
(*Ledaea perditalis*)
aka Buttonbush owlet
13-16 mm.

Thin-lined Owlet
(*Isogona tenuis*)
11-15 mm.

Definite Tussock Moth
(*Orgyia definita*)
30 mm.

Common Fungus Moth
(*Metalectra discali*)
20-29 mm.

Dark-spotted Palthis
(*Palthis angulalis*)
19-23 mm.

Faint-spotted Palthis
(*Palthis asopialis*)
19-23 mm.

Red-lined Panopoda
(*Panopoda rufimargo*)
22-25 mm.

Brown Panopoda
(*Panopoda carneicosta*)
38-46 mm.

Maple Looper
(*Parallelia bistriaris*)
33-43 mm.

Dark-banded Owlet
(*Phalaenophana pyramusalis*)
21-25 mm.

Black-banded Owlet
(*Phalaenostola larentioides*)
17-24 mm.

Speckled Renia Moth
(*Renia adspergillus*)
25 mm.

Discolored Renia Moth
(*Renia discoloralis*)
35-45 mm.

Spotted Grass Moth
(*Rivula propinqualis*)
15-19 mm.

Curved-lined Owlet
(*Phyprosopus callitrichoides*)
28-35 mm.

Deadwood Borer Moth
(*Scolecocampa liburna*)
35-43 mm.

The Herald
(*Scoliopteryx libatrix*)
44 mm.

Six-spotted Gray
(*Spargaloma sexpunctata*)
25-29 mm.

Smoky Tetanolita
(*Tetanolita mynesalis*)
20-25 mm.

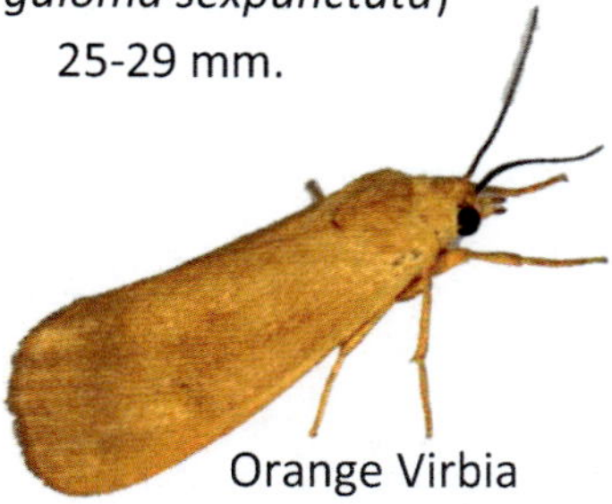

Orange Virbia
(*Virbia aurantiaca*)
22 mm.

Tawny Virbia
(*Virbia opella*)
25 mm.

Maple Zale
(*Zale galbanata*)
42-50 mm.

Lunate Zale
(*Zale lunata*)
40-55 mm.

Grayish Fan-Foot
(*Zanclognatha pedipilalis*)
24-30 mm.

Family Euteliidae - Euteliid Moths

Euteliid moths tend to sit with their wings folded and abdomens bent up, looking much like a dead leaf.

Dark Marathyssa Moth
(*Marathyssa inficita*)
25-28 mm.

Eyed Paectes
(*Paectes oculatrix*)
23-25 mm.

Family Gelecticidae – Twirler Moths

Galacticidae is a recently designated Family of moths previously classified in a number of different families. Reclassification was based on DNA analysis.

Eye-ringed Chionodes Moth
(*Chionodes discoocellella*)
8 mm. TL

Mimosa Webworm Moth
(*Homadaula anisocentra*)
10 mm. TL

White Stripe-backed Moth
(*Arogalea cristifasciella*)
6 mm. TL

Family Geometridae

A large family (~1400 species in North America), Geometers (meaning "measuring the earth") are named for their caterpillars: inchworms. The cats have a limited set of prolegs. With legs only on the front and rear, they have a characteristic "inching" locomotion.

Common Gray
(*Anavitrinella pampinaria*)
23-34 mm.

Blackberry Looper
(*Chlorochlamys chloroleucaria*)
14-23 mm.

Mottled Gray Carpet
(*Cladara limitaria*)
21-28 mm.

Bent-line Carpet
(Costaconvexa centrostrigaria)
male
17-23 mm.

Packard's Wave
(*Cyclophora packardi*)
17-23 mm.

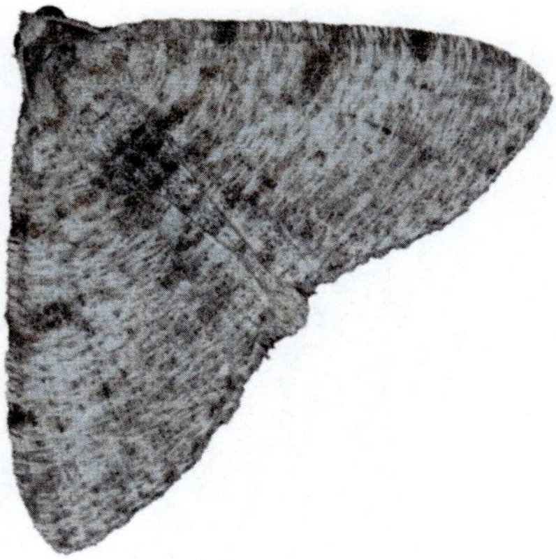

Hollow-spotted Angle Moth
(*Digrammia gnophosaria*)
18-23 mm.

Faint-spotted Angle Moth
(*Digrammia ocellinata*)
21-27 mm.

Somber Carpet
(*Disclisioprocta stellata*)
25-33 mm.

Tulip-tree Beauty
(*Epimecis hortaria*) (M)
43-55 mm.

Small Engrailed Moth/Saddleback Looper
(*Ectropis crepusclaria*)

Deep Yellow Euchlaena
(*Euchlaena amoenaria*)
30-50 mm.

Lesser Grapevine Looper
(*Eulithis diversilineata*)
25 mm

Common Eupithecia
(*Eupithecia miserulata*)

Eupithecia sp.

Confused Eusarca
(*Eusarca confusaria*)
29-41 mm.

Curve-tooth Geometer
(*Eutrapela clemataria*)
38-56 mm.

Fine-lined Gray Moth
(*Exelis pyrolaria*)
22-25 mm.

Texas Grey
(*Glenoides texanaria*)
16-24 mm.

Chickweed Geometer
(*Haematopis grataria*) ♂
20-25 mm.

Three-spotted Fillip
(*Heterophleps triguttaria*)
18-24 mm.

Red-Bordered Wave Moth
(*Idaea demissaria*)
14-19 mm.

Black-dotted Ruddy Moth
(*Ilexia intractata*) ♀
21-31 mm.
Larva eats American holly

Hypagyrtis sp.
likely an Esther Moth
(*Hypagyrtis esther*)
25-45 mm.

One-spotted Variant
(*Hypagyrtis unipunctata*)
20-47 mm.

Umber Moth
(*Hypomecis umbrosaria*)
30 mm.

Brown-shaded Gray
(*Iridopsis defectaria*)
30 mm.

Bent-line Gray
(*Iridopsis larvaria*)
26-36 mm.

White Spring Moth
(*Lomographa vestaliata*)
15-26 mm.

Stained Lophosis (F)
(*Lophosis labeculata*)
14-15 mm.

Minor Angle
(*Macaria minorata*)
21 mm.
Larva feed on pine

Red-headed Inchworm Moth
(*Macaria bisignata*)
25 mm.

Many-lined Angle
(*Macaria multilineata*)
26-28 mm.

Birch Angle (*Macaria notata*) or Common Angle (Macaria aemulataria) . These species look about the same. M. notata is larger (up to 30mm wingspan vs. 22mm ws for M. notata).

Macaria sp
18-35 mm.

Canadian Melanolophia
(*Melanolophia canadaria*)
30-36 mm.

Signate Melanolophia
(*Melanolophia signataria*)
30-36 mm.

<

Red-fringed Emerald
(*Nemoria bistriaria*)
22 mm.
The spring brood of Red-fringed Emeralds is coffee-colored. The summer brood is green.

Red-borded Emerald
(*Nemoria lixaria*)
21-25 mm.
Looks a lot like the Red-fringed Emerald summer brood but has small black discal spots on each wing.

Green Pug
(*Pasiphila rectangulata*)
15-20 mm.

Juniper-twig Geometer
(*Patalene olyzonaria*)
21-25 mm.

Small Phigalia
(*Phigalia strigataria*)
The first moth of the year. Observed 3 Mar while snow was still on the ground but temperature had soared into the 60s. Male wings are 14-18 mm. long; female wings are greatly reduced and do not support flight.

Common Tan Wave
(*Pleuroprucha insulsaria*)
14-21 mm.

Soft-lined Wave
(*Scopula inductata*)
20-25 mm.

The Large Lace Border is a good illustration of diversity within moth species. Each of these moths is *S. limboundata*. The tan version is the "enucleate" form which was formerly considered a separate species.

Large Lace-Border
(*Scopula limboundata*)
25-35 mm.

Wavy-lined Emerald
(*Synchlora aerata*)
17 mm.

Toothed Brown Carpet
(*Xanthorhoe lacustrata*)
20 mm.

False Crocus Geometer
(*Xanthotype urticaria*)
30-40 mm.

Family Heliozelidae – Shield Bearer Moths

Coptodisca cercocarpella
2-3 mm.

Family Lasiocampidae – Lappit Moths

Dot-lined White
(*Artace cribrarius*)
25 mm.

Eastern Tent Caterpillar Moth
(*Malacosoma americana*)
25 mm.

Family Lecithoceridae – Longhorn Moths

Square-spotted Martyringa
(*Martyringa latipennis*)
25 mm.

Family Limacodidae – Slug Moths

The larvae of slug moths move like slugs, often having urticating spines.

Saddleback Caterpillar
(*Acharia stimulea*)

Purple-crested Slug Moth
(*Adoneta spinuloides*)
9 mm.

Yellow-collared Slug Moth
(*Apoda y-inversum*)
aka Inverted Y Slug Moth
21-30 mm.

Shagreened Slug Moth
(*Apoda biguttata*)
9 mm.

Spun Glass Slug Moth
(*Isochaetes-beutenmuelleri*)
10 mm.

Yellow-shouldered Slug Moth
(*Lithacodes fasciola*)
10-13 mm.
aka Ochre-winged Hag Moth

Smaller Parasa
(*Parsa chloris*)
8 mm.

Stinging Rose Caterpillar Moth
(*Parasa indetermina*))
8 mm.

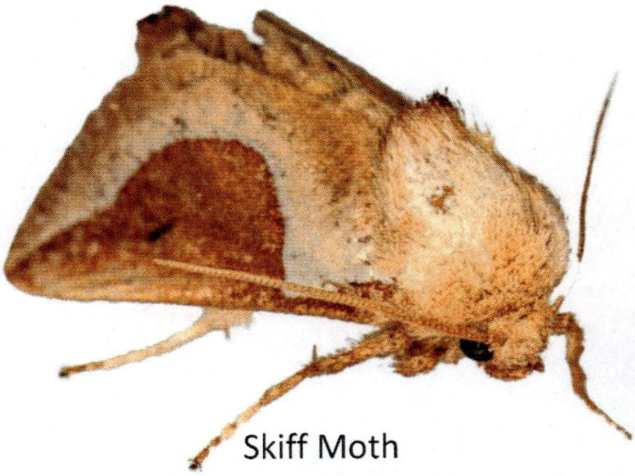

Skiff Moth
(*Prolimacodes badia*)
24-35 mm.

Early Button Slug Moth
(*Tortricidia testacea*)
12 mm.

Family Megalopygidae – New World Flannel Moths

White Flannel
(*Norape ovina*)
24 mm.

Black-waved Flannel
(*Megalopyge crispata*)
24 mm.

Family Mimallonidae – Sack-bearer Moths

Mimallonids are limited to the New World, but of the estimated 300 species, only five are found in the United States. The larval Mimalloids create protective "cases" of leaves, silk, and frass.

Scalloped Sack-bearer
(*Lacosoma chiridota*)
24-35 mm.

Family Noctuidae – Owlet Moths

Afflicted Dagger
(*Acronicta afflicta*)
35 mm.

American Dagger
(*Acronicta americana*)

Cherry Dagger
(*Acronicta hasta*)
35-45 mm.
Aka Speared Dagger

Eclipsed Oak Dagger
(*Acronicta increta*)
22 mm.

Greater Oak Dagger
(*Acronicta lobeliae*)
42-60 mm

Interrupted Dagger
(*Acronicta interrupta*)
34-42 mm.

Long-winged Dagger
(*Acronicta longa*)

Smeared Dagger
(*Acronicta oblinita*)
35-50 mm.

Maple Dagger
(*Acronicta retardata*)
25-32 mm.

Ruddy (aka Hackberry) Dagger Moth & Cat
(*Acronicta rubricoma*)
38-44 mm.

Ipsilon Dart
(*Agrotis ipsilon*)
38-48 mm.

Unspotted Looper
(*Allagrapha aerea*)
28-42 mm.

Copper Underwing
(*Amphipyra pyramidoides*)
38-52 mm

Green Cutworm
(*Anicla infecta*)
30-35 mm.

Snowy Dart
(*Anicla illapsa*) or *(Euagrotis illapsa)*
29-35 mm.

The Slowpoke
(*Athetis tarda*)
23-35 mm.

Common Looper Moth
(*Autographa precationis*)
23-35 mm.

Cattail Borer Moth
(*Bellura obliqua*)
24 mm.

Reddish Speckled Dart
(*Cerastis tenebrifera)*
30 mm.

An early season moth, this individual was seen in mid-March.

Tobacco Budworm Moth
(*Chloridea virescens)*
28-35 mm.

Bent-line Dart
(Choephora *fungorum)*

Soybean Looper
(*Chrysodeixis includens)*
28-39 mm.

This pupa was found in our garden. The two "adult" pictures were taken the day it emerged.

Cloaked Marvel
(*Chytonix palliatricula*)
29-35 mm.

Dusky Groundling
(*Condica vecors*)
30-38 mm.

American Bird's-Wing
(*Dypterygia rozmani*
34-39 mm.

Beautiful Wood-Nymph
(*Eudryas grata*)
35-38 mm.

Grateful Midget
(*Elaphria grata*)
20-26 mm.

Verigated Midget
(*Elaphria versicolor*)
22 mm.

Master's Dart
(*Feltia herilis*)
34-44 mm.

Corn Earworm
(*Helicoverpa zea*)
32-45 mm.

The Wedgeling
(*Galgula partita*)
20-26 mm.

Black Wedge-spot
(*Homophoberia apicosa*)
14 mm. body length

Explicit Arches
(*Lacinipolia explicata*)
15-17 mm. body length

Bristly Cutworm Moth
(*Lacinipolia renigera*)
14-15 mm.

Linda's Wainscot
(*Leucania linda*)
20-21 mm.

Phragmites Wainscot Moth
(*Leucania phragmitidicola*) caterpillar
25 mm.

False Wainscot
Prob. (*Leucania pseudargyria*)
26 mm.

Small Mossy Glyph or Lithacodia
(*Lithacodia musta*)
15-19 mm.

Ursula Wainscot
(*Leucania ursula*)
35 mm.

Black-dotted Glyph
(*Maliattha synochitis*)
10 mm. body length
AKA Black-dotted Maliattha or
Black-dotted Lithacodia

Green Leuconycta
(*Lithacodia diphteroides*)
15-19 mm.
This is the rare white form of a moth
that is usually predominantly green.

Black-bordered Lemon
(*Marimatha nigrofimbria*)
18-22 mm.

The White-speck
(*Mythimna unipuncta*)
a.k.a. Armyworm moth
35-47 mm.

Bronzed cutworm
(*Nephelodes minians*)
15-20 mm

Common Pinkband
(*Ogdoconta cinereola*)
10-15 mm.

White-marked Tussock
(*Orgyia leucostigma*)
10-15 mm.

Aerial Brown
(*Ozarba aeria*)
33-50 mm.

Eastern Panthea
(*Panthea furcilla*)
33-50 mm WS

Red Groundling Moth
(*Perigea xanthioides*)
29 mm.

Spotted Phosphila
(*Phosphila miselioides*)
30 mm.

Hebrew Moth
(*Polygrammate hebraeicum*)
29 mm.

Large Mossy Glyph
(*Protodeltote muscosula*)
30 mm

Arcigera Flower Moth
(*Schinia arcigera*)
22-25 mm.

Lynx Flower Moth
(*Schinia lynx)*
22-25 mm.

Ragweed Flower Moth
(*Schinia rivulosa*)
25-31 mm.

Yellow-striped Armyworm Moth
(*Spodoptera ornithogalli*)
32-44 mm.

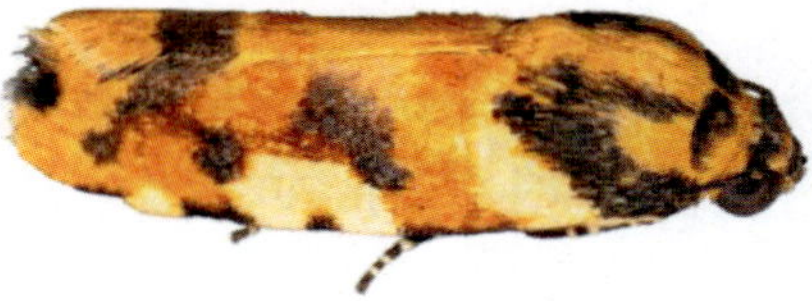

Southern Spragueia Moth
(*Spragueia dama*)
16-22 mm.

Common Spragueia (moth & cat)
(*Spragueia leo*)
12-18 mm.

FEMALE
Exposed Bird Dropping Moth
(*Tarache aprica*)
15-29 mm
MALE

Greater Black-letter Dart
(*Xestia dolosa*)
40 mm.

Family Nolidae – Nolid Moths

Kin to the Noctuidae, the Nolid moths are also called Tuft moths due to the presence of tufted scales. Most of this family is found in Europe.

Small Baileya
(*Baileya australis*)
21-28 mm.

Sleeping Baileya
(*Baileya dormitans*)
25-31 mm.

Sorghum Webworm
(*Nola cereella*)
12-18 mm.

Coastal Plain Meganola Moth
(*Meganola phylla*)
18-24 mm.

Grey Midget Moth
(Nycteola cinereana)
12-14 mm.

Forgotten Frigid Owlet
(*Nycteola metaspilella*)
23 mm.

Family Notodontidae – Prominent Moths

Black-etched Prominent
(*Cerura scitiscripta*)

Apical Prominent
(*Cerura apicalis*)
20 mm.

Angle-lined Prominent
(*Clostera inclusa*)
25-32 mm.

Walnut Caterpillar Moth
(*Datana integerrima*)
35-50 mm.

Spotted Datana
(*Datana perspicua*)
45-56 mm.

Wavy-lined Heterocampa
(*Heterocampa biundata*)
25-30 mm. TL

Common Gluphisia
(*Gluphisia septentrionis*)
25-33 mm.

Saddled Prominent
(Heterocampa guttivitta)
40 mm. (23-25 mm TL)

This small caterpillar will change appearance five time as it molts and grows. The first instar (orange with antlers) is 5 mm. The 14 mm green version is the same caterpillar 10-days later.

Oblique Heterocampa
(*Heterocampa obliqua*)
44 mm.
Eats oak leaves

Small Heterocampa
(*Heterocampa subrotata*)
16-20 mm. TL

White-dotted Prominent
(*Nadata gibbosa*)
50 mm.

Variable Oakleaf Caterpillar Moth
(*Lochmaeus manteo*)
37-50 mm.

Unicorn Prominent
(*Schizura unicornis*)
24-36 mm.

Black-blotched Schizura
(*Schizura leptinoides*)

Family Oecophoridae – Concealer Moths

Suzuki's Promalactis
(*Promalactis suzukiella*)
5 mm.

A native to Japan, Korea, and Tiawan, *P. suzukiella* was introduced to the U.S. in the early 2000s and is now established throughout the eastern U.S.

Family Opostegidae -- White Eye-capped Moths

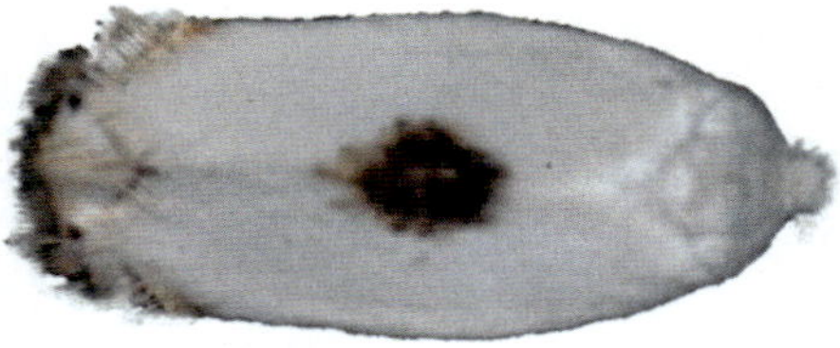

Gooseberry Barkminer
(*Pseudopostega quadristrigella*)
5-6 mm. TL

Family Psychidae – Bagworms

Psychidae sp .

Family Pterophoridae – Plume Moths

Morning-glory Plume Moth
(*Emmelina monodactyla*)
18-27 mm.

Family Pyralidae – Pyralid (or Snout) Moths

A large family of small moths; almost 700 species in North America.

Tricolored Acrobasis
(*Acrobasis tricolorella*)
10 mm. TL

Pink-masked Pyralid Moth
(*Aglossa disciferalis*)
18 mm.

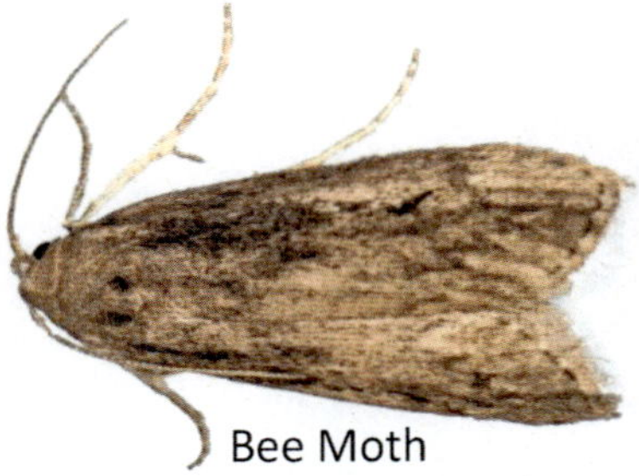

Bee Moth
(*Aphomia terrenella*)
10 mm. TL

Posturing Arta
(*Arta statalis*)
5 mm. TL

Dimorphic Macalla
(*Epipaschia superatalis*)
10 mm. TL

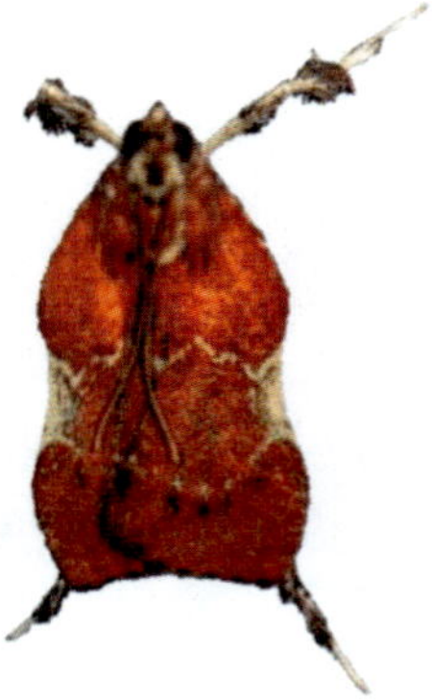

Boxwood Leaftier Moth
(*Galasa nigrinodis*)
6 mm. TL

American Plum Borer
(*Euzophera sedmifuneralis*)
10 mm. TL

Clover Hayworm
(*Hypsopygia costalis*)
16-24 mm.

Broad-banded Eulogia
(*Eulogia ochrifrontella*)
11-15 mm.
The only species in the Eulogia genus

Yellow-fringed Dolichomia
(*Hypsopygia olinalis*)
13-19 mm.

Maple Webworm
(*Pococera asperatella*)
14 mm. TL

Pococera sp.
possibly Striped Oak Webworm
(*Pococera expandens*)
26 mm.
a.k.a. Double-humped Pococera Moth

Indian Meal Moth
(*Plodia interpunctella*)
13-20 mm.

Sweetgum Leafroller Moth
(*Sciota uvinella*)
~12 mm.

Dimorphic Tosale
(*Tosale oviplagalis*) male
15 mm. body length

Family Saturniidae – Giant Silkworm Moths

North American Luna Moth
(*Actias luna*)
120 mm.

Spicebush Silkmoth
(*Callosamia promethea)*
100 mm.

Rosy Maple Moth
(*Dryocampa rubicunda*)
50 mm.

Imperial Moth
(*Eacles imperialis*)
80-175 mm..

Cecropia Moth caterpillar
(*Hyalophora cecropia*)

Family Sesiidae – Clearwing Moths

Squash Vine Borer Moth
(*Melittia cucurbitae*)
28-32 mm.

Family Sphingidae – Sphinx Moth

Elm Sphinx
(*Ceratomia amyntor*)
82-115 mm.

Catalpa Sphinx
(*Ceratomia catalpae*)
65-95 mm.

Virginia Creeper Sphinx
(*Darapsa Myron*
45-65 mm.

Pandorus Sphinx
(*Eumorpha pandorus)*
87-115 mm.

Snowberry Clearwing
(*Hemaris diffinis*)
35-50 mm.

Looking and sounding like a hummingbird, these moths are one of the few species that are active during the day.

Tobacco Hornworm
(*Manduca sexta*)
Looks and eats like the tomato hornworm but has a red "horn" while the Tomato Hornworm has a black "horn."

Hornworms are often parasitized by small wasps who lay eggs just under the skin of the hornworm. The hornworm will generally die soon after the wasps hatch.

Five-spotted Hawk Moth
(*Manduca quinquemaculatus*)
90-135 mm
Adult form of the Tomato Hornworm

Blinded Sphinx
(*Paonias excaecatus*)
55-95 mm.

Small-eyed Sphinx
(*Paonias myops*)
45-75 mm.

White-lined Sphinx adult & caterpillar
(*Hyles lineata*)
60-90 mm.

Abbott's Sphinx
(*Sphecodina abbottii*)
51-70 mm.

Factoid: Some Native Americans skewered and roasted *Hyles lineata* caterpillars as snacks.

Family Tineidae – Fungus & Clothes Moths

Eastern Grass Tubeworm Moth
(*Acrolophus plumifrontella*)
28 mm.
Some sources place this moth in its own Family (Acrolophidae)

Clemens' Bark Moth
(*Xylesthia pruniramiella*)
Aka Speckled Xylesthia
13 mm.

Family Tortricidae – Toricid Moths (Leafrollers)

Omniverous Leafroller
(*Archips purpurana*)
16-25 mm.

Yellow-winged Oak Leafroller
(*Argyrotaenia quercifoliana)*
16-24 mm.

Red-banded Leafroller
(*Argyrotaenia velutinana)*
13-20 mm

Spotted Fireworm
(Choristoneura parpllela)
21-23 mm.

Choristoneura obsoletana
21-24 mm.

Oblique-banded Leafroller
(*Choristoneura rosaceana*)
8-11 mm.

Garden Tortrix
(*Clepsis peritana*)
10-15 mm.

The Batman
(*Coelostathma discopunctana*)
11-15 mm.

Verbena Bud Moth
(*Endothenia hebesana*)
11-15 mm.

Pecan Bud Moths
(*Gretchena bolliana*)
17 mm.

Ragweed Borer
(*Epiblema strenuana*)
15 mm.

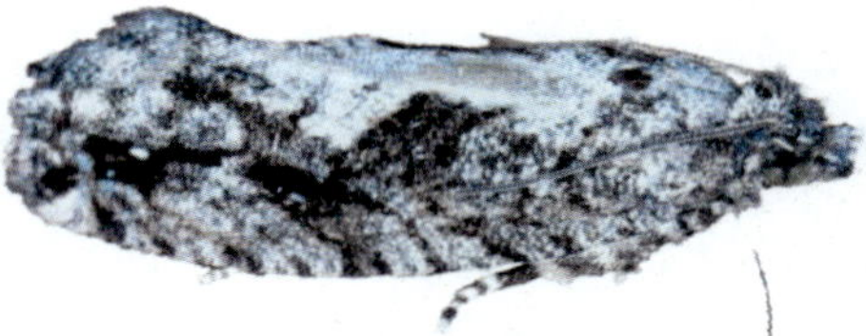

Dotted Gymnandrosoma Moth
(*Gymnandrosoma punctidiscanum*)
20 mm.

The Astronomer
(*Olethreutes astrologana*)
14-19 mm.

Black-shaded Platynota
(*Platynota flavedana*)
35-42 mm.

Maple Twig Borer
(*Proteoteras aesculana*)
11-18 mm.

Tufted Apple Bud Moth
(*Platynota idaeusalis*)
12-25 mm.

Sparganothis Fruitworm
(*Sparganothis sulfureana*)
10-20 mm.

Order Mantodea – Mantids

Family Mantidea – Mantis

Everybody likes the praying mantis, right? They're large, tolerant of being handled, and they eat garden pests. They're also pious. Well, not so fast. They are indeed large and tolerant, but as far as their eating habits, they're pretty much indiscriminate. They'll eat anything they can catch, including native pollinators and each other.

Carolina Mantis
(*Stagmomantis carolina*)
Up to 60 mm. (males somewhat smaller)

There are at least four species of mantis in much of the United States, but in Virginia you will find mostly the native Carolina mantis (*Stagmomantis Carolina*) and the imported Chinese mantis (*Tenodera sinensis*). Brought in for "pest control," *T. sinensis* is larger than our native mantis and being bigger, often eats it.

Identification can be tricky. Both species exhibit a range of coloration from green to brown to gray. The female Chinese mantis is about 4" long, an inch longer than the full grown Carolina mantis, but for mantids less than 3" you'll have to look closely. The female Chinese mantis has wings that extend beyond her abdomen, while the Carolina mantis has shorter wings. The face of the Chinese mantis has vertical striping; legs of the Carolina mantis are generally blotchy. Two other imports that look similar are the Narrow-winged and the European mantis. If you find a mantis with a bright orange spot on its chest, you've got a Narrow-winged mantis (*Tenodera angustipennis);* a mantis with a black-and-white mark on its foreleg is a European mantis (*Mantis religiosa*). I have not observed either of these species.

Chinese Mantis
(*Tenodera sinensis*)
Up to 110 mm., the largest mantid in North America

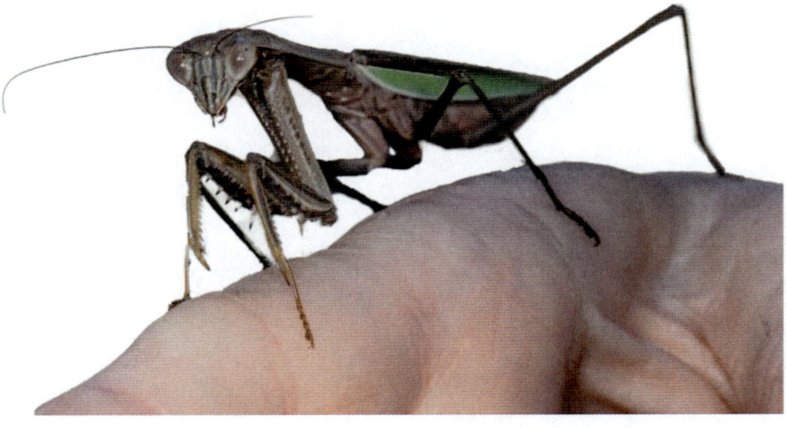

Order Megaloptera – Alder/Dobson/Fishflies

Family Chauliodinae – Fishflies

Fishflies are an indicator species requiring fresh, clear water to thrive. Pollution has caused the fishfly to disappear from much of its natural habitat. Although this species can live several years, most of that time is spent as an aquatic nymph. The adult fishfly will live only about a week.

Spring Fishfly
(Chauliodes rastricornis)
40 mm.

Eastern Dobsonfly
(Corydalus cornutus)
140 mm.

Neohermes concolor
40 mm.

Oder Microcoryphia – Bristletails

The bristletails have been around a long time. They evolved about the same time as the spider, around 400,000,000 years ago. Bristletails are most notable for the three long tail-like projections: a long epiproct with shorter cerci on each side. This individual was under a log, a common location for bristletails. If disturbed, this 8 mm insect uses its tail and abdominal structures to propel itself up to 300 mm in the air. Having thin "skin", it must remain in a moist environment to avoid dehydration. It may live up to four years on a diet of fungi, moss, and algae.

Microcoryphia sp
8 mm.

Order Neuroptera – Lacewings and Antlions

Family Chrysopidae

Neuropterans are characterized by four wings about the same size, containing distinct veining. Common in Virginia gardens are the green, golden-eyed, and brown lacewings. While the adults feed on algae, lichens, moss, and detritus, the nymphs are predators of numerous garden pests, with a particular appetite for aphids.

A Green Lacewing
(Chrysopidae sp.)
One of 84 species in NA

Green Lacewing nymph
As the nymph hunts it will armor itself with the remains of its victims.

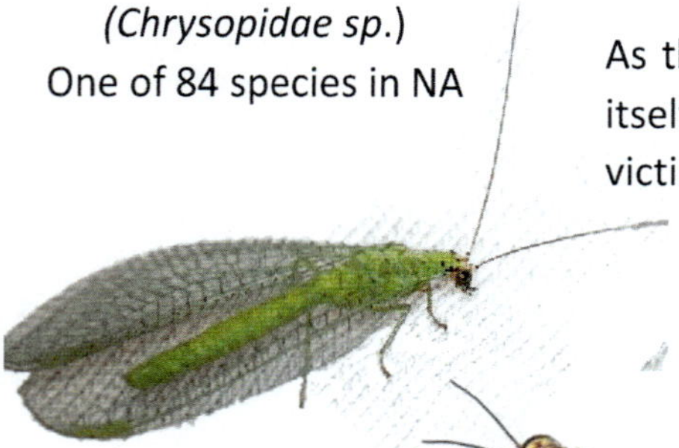

Golden-eyed Lacewing
(*Chrysopa oculata)*
18 mm.

Say's Mantidfly
(*Dicromantispa sayi)*
15 mm.

Family Hemerobiidae – Brown lacewings

Brown Lacewing
(*Hemerobiidae sp.)*
8 mm.

Order Odonata – Dragonflies & Damselflies

Merciless predators and skillful aviators, the odonata bring drama and beauty to our ponds and fields. Similar in form and function, dragonflies and damselflies can be separated by observing them at rest. Dragonflies perch with wings outstretched or held up; damselflies are smaller and hold their wings along their bodies. Dragonflies have large eyes that touch; damselflies have smaller eyes that are well separated.

Family Aeshnidae – Darners

Common Green Darner
(*Anax junius*)
70-80 mm.

Comet Darner
(Anax longipes)
78 mm.

Family Calopterygidae – Damselflies

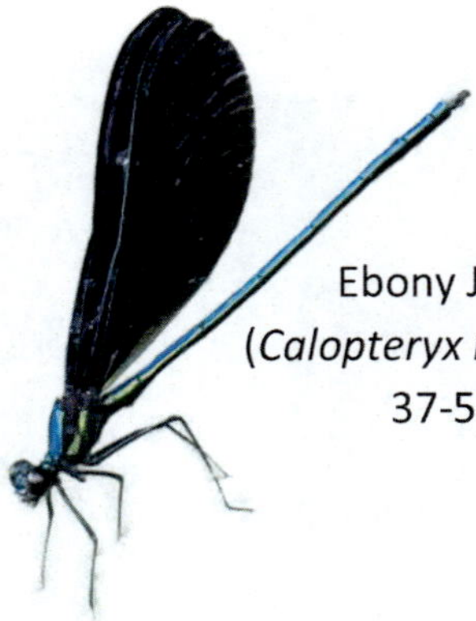

Ebony Jewlwing
(*Calopteryx maculata*) ♂
37-57 mm.

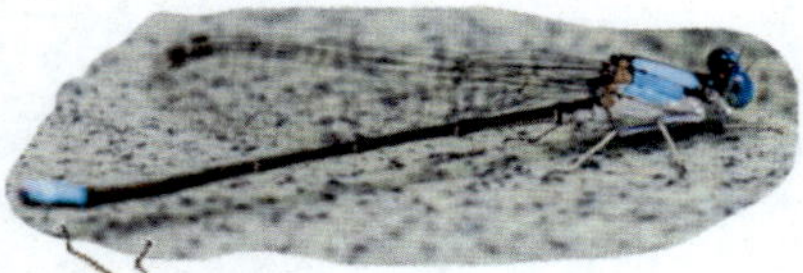

Blue-fronted Dancer
(*Argia apicalis*) ♂
33-40 mm.

Note: The female A. apicalis comes in both blue and brown versions, making identification difficult

American Bluet
(*Enallagma* sp.) ♂
35 species in North America

Fragile Forktail
(*Ischnura posita*) -
37-57 mm.

Family Gomphidae – Clubtails

Lancet Clubtail
(*Phanogomphus exilis*)
40 mm.

Family Libellulidae – Skimmers

Calico Pennant
(*Celithemis elisa*) ♂
30 mm.

Males are not territorial; perch along edges of ponds, on the lookout for incoming females. Pairs perch on vegetation near water while mating, oviposit as a pair.

Eastern Pondhawk
(*Erythemis simplicicollis*)
36-46 mm.

While females and young males maintain this lovely green color, adult males turn blue, retaining the green only on their faces.

Male -- Common Whitetail Skimmer -- Female
(*Plathemis lidia*)
50 mm.

Slaty Skimmer
(***Libellula incesta***)
53 mm.

Widow Skimmer
(*Libellula luctuosa*) ♂
24-32 mm.

Spangled Skimmer
(*Libellula cyanea*)
40-48 mm.

Blue Dasher
(*Pachydiplax longipennis)*
25-43 mm.

Eastern Amberwing
(*Perithemis tenera*)
21-24 mm.

Order Orthoptera – Grasshoppers/Crickets/Katydids

Stridulation can be fun! Although many animals stridulate (make sound by rubbing two body parts together), it is the crickets, katydids, and grasshoppers that are best known for their sounds. These orthopterans create a symphony on a summer's evening. The other stridulators are much quieter. If you are listening to the stridulation of a bark beetle for instance, you are probably too close to your work.

Family Acrididae – Grasshoppers

Differential Grasshopper
(*Melanoplus differentialis*)
Males: 28-37mm, Females 34-50mm

Cattail Toothpick Grasshopper
(*Leptysma marginicollis marginicollis*)
Males: 28-31 mm., females: 31-38 mm.

Green-legged Spur-throat Grasshopper
(*Melanoplus viridipes*)
Males: 27-31 mm, females: 37-45 mm

Olive-green Swamp Grasshopper
(*Paroxya clavuliger*)
Males: 20-33 mm, female: 29-46 mm

American Bird Grasshopper
(*Schistocerca americana*)
40-70 mm.

Family Gryllacrididae – Raspy Crickets

Carolina Leaf-roller
(*Camptonotus carolinensis*)
12-15 mm.

Bites through leaf to form a flap. The flap is folded over, the edge is pulled down, edges are then glued together with silk from a gland on the mouth. Sometimes uses the pods of Bladdernut (*Staphylea trifolia*) as a shelter instead of a leaf

Family Gryllidae – True Crickets

Robust Ground Crickets
(Allonemobius sp.)

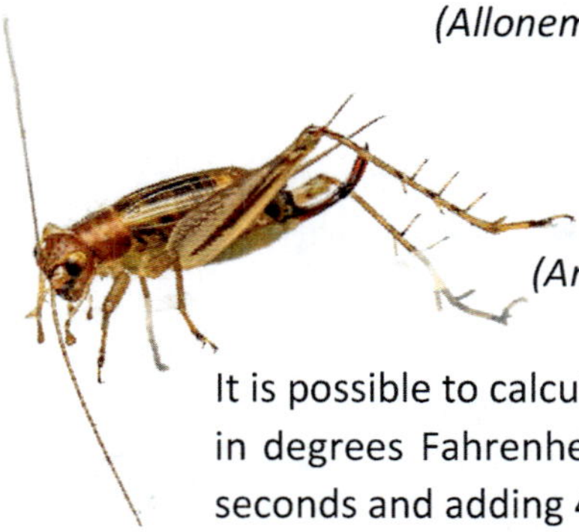

Say's Trig
(Anaxipha exigua)
12 mm.

It is possible to calculate approximate air temperature in degrees Fahrenheit by counting the chirps per 14 seconds and adding 40.

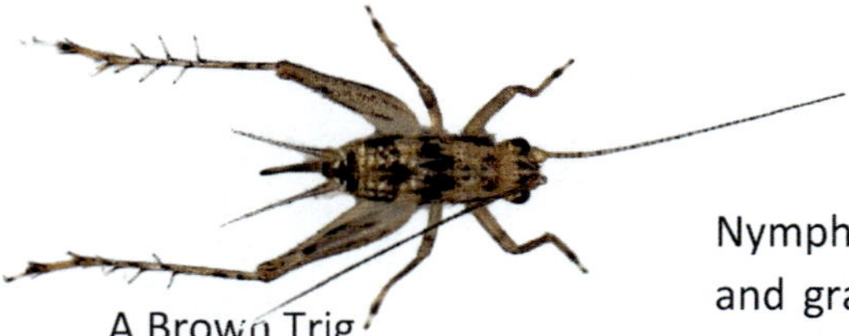

A Brown Trig
(Anaxipha sp.)

Tree Crickets
(*Oecanthus sp.)*
4-6 mm.

Nymphs resemble small adults and gradually develop external wing buds. They live in the same habitat as adults, typically taking the same food.

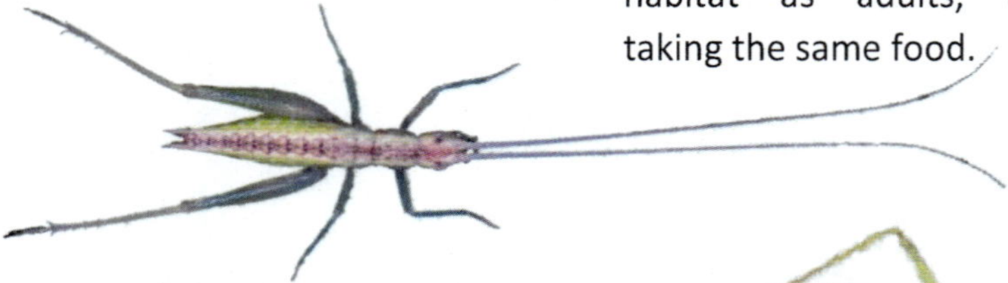

Broad-winged Tree Cricket
(*Oecanthus latipennis*)
4-6 mm.

Family Rhaphidophotidae – Crickets

-

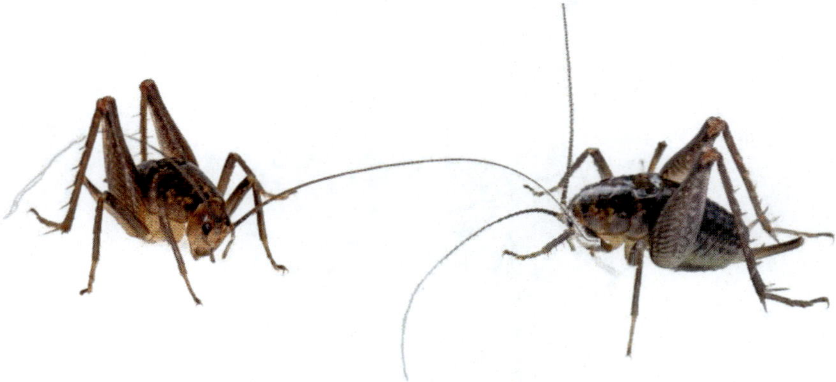

Camel Cricket
(*Ceuthophilus sp.*) ♀

One of about 90 species. The camel cricket is frequently found in residential basement, crawl spaces, and similar environments. Omniverous.

Greenhouse Camel Cricket
(*Diestrammena asynamora*}
11-14 mm.

Family Tettigoniidae – Katydids

Lesser Meadow Katydids
(Conocephalus sp.)
12 mm.

Short-winged Meadow Katydid
(*Conocephalus brevipennis*)
12 mm.

Slender Meadow Katydid
(*Conocephalus fasciatus*)
25 mm.

Lesser Angle-winged Katydid
(*Microcentrum retinerve*)
44-53 mm.

Greater Angle-wing Katydid
(*Microcentrum rhombifolium*)
52-63 mm.

The skull of the angle-wing katydid looks disturbingly like that of people.

← Front Back →

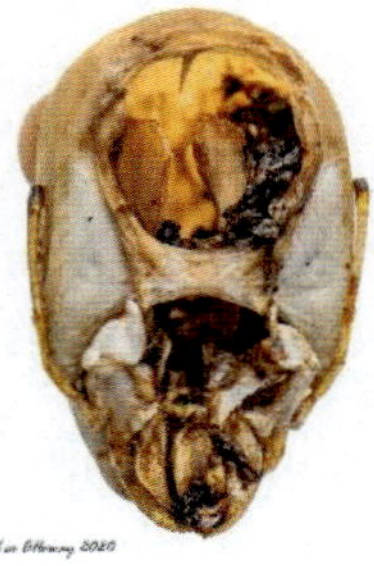

Modest Katydid
(*Montezumina modesta*) ♂
50 mm.

Common Conehead
(*Neoconocephalus sp*)
12 mm.

Lesser Pine Katydid
(*Orchelimum minor*)
30 mm.

Fork-tailed Bush Katydid nymph
(*Scudderia furcata*)
45 mm

Bush Katydid nymph
(*Scudderia sp*)

Order Plecoptera – Stoneflies

Family Pteronarcyidae

The stonefly lives most of its life as an aquatic nymph (naiad). After several years and many molts as a naiad it will emerge as an adult for a few short weeks. During that time some species will not eat, although some are plant eaters. Stoneflies require well-oxygenated water to live, so are considered an indicator of a good water quality.

American Salmonfly
(Pteronarcys dorsata)
31-48 mm.
or
Pteronarcys pictetii

Plecoptera sp

Order Psocoptera – Barklice

Family Amphipsocidae – Hairy-winged Barklice

The barklice are an ancient order with about 5500 identified species worldwide. The order is also known as booklice and includes species that are pests in grain storage and yes, some that feed on the paste in books. The following species, however, were found outside feeding on moss, algae, and detritus.

Polypsocus corruptus

4 mm.

Family Dasydemellidae – Shaggy Barklice

Teliapsocus conterminus
~3 mm

Often associated with birch trees, this individual was nowhere near a birch. Found in the field. *T. conterminus* is the only shaggy psocid in our area.

Family Psocidae – Common Barklice

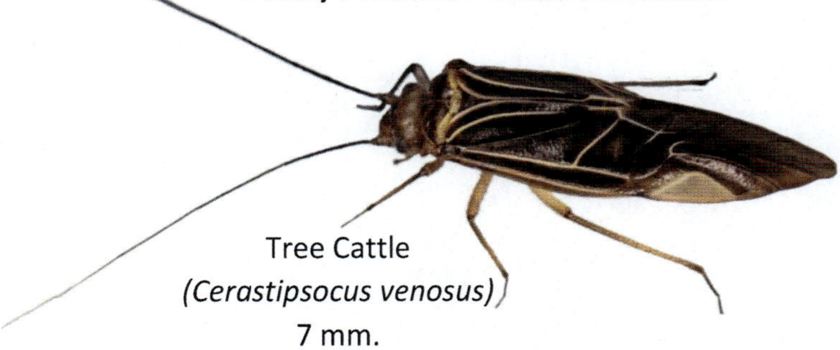

Tree Cattle
(Cerastipsocus venosus)
7 mm.

Family Stenopsocidae – Narrow Barklice

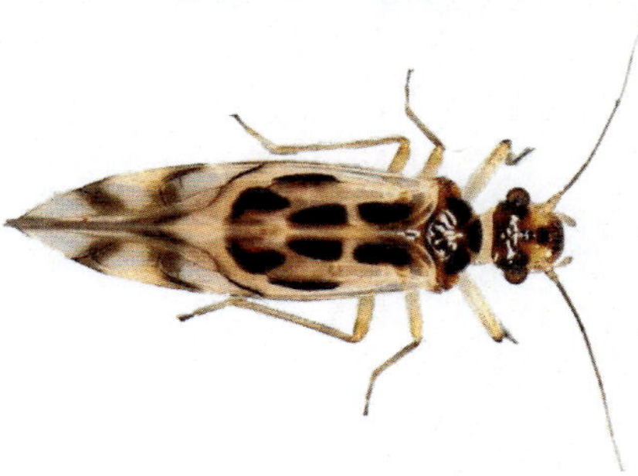

Graphopsocus cruciatus
4 mm.

An Asian import in 1930, *G. cruciatus* has spread throughout the East and West coasts and is now moving inland.

Order Thysanoptera – Thrips

Small (most are about 1mm) and annoying, thrips are generally something you don't want around. Although these critters may be predators of mites, thrips seem to contribute little and cause significant harm to plants. If they establish themselves in your greenhouse or houseplants, they can rapidly multiply and move into your household items or your computer screen. With 6,000 species identified, thrips are clearly an evolutionary success story.

Family Phlaeothripidae

Phlaeothripidae sp
2 mm.

Order Trichoptera – Caddisflies

Related to the moths and butterflies, caddisflies have hairy, membranous wings and aquatic larvae. Like the stoneflies, caddisflies spend most of their lives as naiads, living for only a few weeks in their adult form. Life is perilous for the caddisfly. The naiads are eaten by fish and other aquatic predators, and the adults are eaten by fish, bats, birds, spiders, and various insects. The newly-eclosed adult caddisfly is particularly attractive to fish, so many fishing lures are modeled after caddisflies.

Family Hydroptilidae

A Microcaddisfly
4 mm

Family Phryganeidae

A Giant Casemaker Caddisfly
19 mm

Family Lepidostomatidae

Lepidostoma sp
A Bizarre Caddisfly
20 mm

Family Leptoceridae – Long-horned Caddisflies

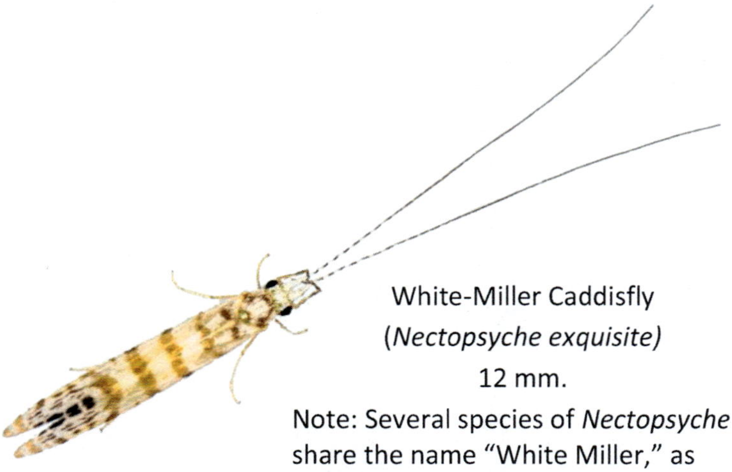

White-Miller Caddisfly
(*Nectopsyche exquisite)*
12 mm.

Note: Several species of *Nectopsyche* share the name "White Miller," as does the genus in general.

Unidentified caddisfly
7mm.

Unidentified caddisfly

Caddisfly gelatenous egg mass – 4 mm.

Non-Insects That Look Like Insects

Class Entognatha

Order Entomobryomorpha – Elongate-bodied Springtails (Collembola)

Although collembola may be as big as 6mm (1/4-inch), most are less than 2mm. They are abundant in garden soil but rarely seen. Their abundance is staggering, with estimates of 100,000 individuals in each cubic meter of earth. Having no external mouth parts, they employ niche strategies to obtain food, some species feeding on fungi, some "cultivating" groups of micro-organisms for food. Most species have a spring-like "furcula" under the abdomen with which they can propel themselves upward, hence the name "springtail."

Family Entomobryidae – Slender Springtails

Homidia socia

Homidia sauteri

Lepidocyrtus violaceus

Family Hypogastruridae

Snow Flea
(*Hypogastrura harveyi*)
2 mm.

This fascinating springtail is quite cold-tolerant. It can live in snow and will sometimes thrive there to the point that is makes the snow look sooty.

Family Isotomidae

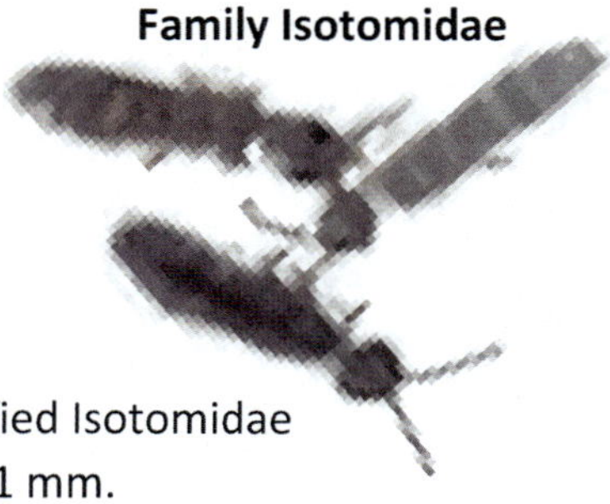

Unidentified Isotomidae
1 mm.

Family Neanuridae

Barber's Springtail
(*Sensillanura barberi*)
2 mm.

Vitronura giselae
2 mm.

Class Malacostraca
Order Isopoda (Suborder Oniscidea)

At first glance these Isopods look like bugs, but they are crustaceans related to crabs, lobsters and the like. While many isopods are aquatic, you will find many living in a cool, moist part of your yard. Some species will roll themselves into small, armored balls when threatened, so are known and loved as "rolly-pollies" or "woodlebugs".

Family Philosciidae – Woodlice

Philoscia muscorum
10 mm.

Family Trachelipodidae– Woodlice

Trachelipus rathkii
14 mm.

References

Excellent reference material is available in print and on the Internet. These are some of my favorites:

Print

Evans, Arthur V., **Beetles of Eastern North America**, Princeton University Press, Princeton 2014. This is my go-to reference for beetles.

Beadle, D. and Leckie, S., **Peterson Field Guide to Moths of Northeastern North America**, Houghton Mifflin Harcort, Boston 2012

Opler, P. and Malikul, V., **Peterson Field Guides -- Eastern Butterflies**, Houghton Mifflin Harcort, Boston 1998

Eaton, Eric R. and Kaufman, Kenn, **Kaufman Field Guide to Insects of North America**, Houghton Mifflin, New York, 2006

Wagner, David L., **Caterpillars of Eastern North America**, Princeton University Press, Princeton, 2005

Gullan, P.J. and Cranston, P.S., **The Insects – *An Outline of Entomology***, Wiley Blackwell, Chichester, 2014 – If you slept through your college entomology class, this is what you missed.

Walters, Martin, **The Illustrated World Encyclopedia of Insects**, Lorenz Books, London 2017

James, David G., **The Book of Caterpillars**, Ivy Press, London, 2017

Marshall, Stephen A., Beetles – **The Natural History and Diversity of Coleoptera**, Firefly Books, Buffalo 2018. Huge, comprehensive, and informative, but not as accessible as Evans' *Beetles of Eastern North America* for our local beetles.

Internet

BugGuide (bugguide.net) –Post an image of your unknown bug and the entomologists, grad students and assorted experts who support BugGuide will identify it for you, often to the species level. Response is generally timely (within one or two days). If you already know the bug you can enter the name and be directed to pictures and information on the species.

INaturalist (inaturalist.org) Another place you can upload images and get community help with identification. Unlike BugGuide, INaturalist maintains an image recognition algorithm that will suggest a tentative ID pending human confirmation. A useful feature is graphic displays of when and where sightings were reported.

Prince William Conservation Alliance: Moths
(http://www.pwconserve.org/wildlife/insects/moths/index.htm)

Ground Beetles – Tribe Harpalini
(https://www.pbase.com/tmurray74/ground_beetles_tribe_harpalini)
Lots of ground beetles look about alike. Photos on this site are not diagnostic but might help steer you toward an ID.

Spiders in Virginia (https://spiderid.com/locations/united-states/virginia/) 57 species of spiders found in Virginia, with good pictures.

Crambids
http://capemaywildlife.com/_templates/mothgroup_crambidae.html)
Good pictures to help identify members of this large family of moths.

Virginia Dragonflies (https://www.inaturalist.org/guides/6115) One of several guides in INaturalist

Virginia Beetles (https://www.beetleidentification.org/beetles-by-state-listing.php?reach=Virginia) Good visual guide to our local beetles

Virginia Insects (https://www.insectidentification.org/insects-by-state.php?thisState=Virginia) An excellent visual guide to more than 1,000 insects found in Virginia

Google (google.com) Google, of course, will find information and images of whatever bug interests you. Often Wikipedia is the first hit.

Index

AA Binomial Nomenclature

ABOUT THE AUTHOR

Educated as a chemist, Lee Alloway was an intern at NASA, a military pilot, and an advisor to the Hungarian Ministry of Defense before serving fourteen years as a civil servant in the Pentagon. He got his first camera when he was five and has documented most of his travels through more than 100 countries. His interest in macro photography led to a passion to study the insects and other arthropods that live unnoticed around us. For the past decade he has traveled around the world photographing bugs, often in the company of entomologists and other macro photographers. He is the author of eight books of poetry and a large collection of seldom heard songs. This is his first book of non-fiction.

Made in United States
Orlando, FL
04 November 2026

90200388R00190